The HIDDEN LIVES *of* OWLS

The HIDDEN LIVES *of* OWLS

The Science and Spirit of Nature's Most Elusive Birds

Leigh Calvez

Illustrations by Tony Angell

SASQUATCH BOOKS
SEATTLE

For Mom and Dad

Printed in the United States of America

Published by Sasquatch Books

20 19 18 17 16 9 8 7 6 5 4 3 2 1

Editor: Gary Luke
Production editor: Em Gale
Design: Joyce Hwang
Illustrations: Tony Angell
Copyeditor: Elizabeth Johnson

Library of Congress Cataloging-in-Publication Data is available.

ISBN: 978-1-63217-025-5

Sasquatch Books
1904 Third Avenue, Suite 710
Seattle, WA 98101
(206) 467-4300
www.sasquatchbooks.com
custserv@sasquatchbooks.com

SFI label applies to text stock only

Author's Note

In keeping with the customary rule established by the International Ornithologist's Union, all official bird names are capitalized.

Contents

A Comparison of Owls by Size ix
Introduction xi

Neighbors: NORTHERN SAW-WHET OWLS 1

Flames of the Forest: FLAMMULATED OWLS 25

Ukpik: SNOWY OWLS 41

Rarely Spotted: NORTHERN SPOTTED OWLS 73

Opportunistic: BARRED OWLS 87

Underground: BURROWING OWLS 101

Synchronicity: NORTHERN PYGMY OWLS 125

The Long and Short of Eide Road: LONG-EARED AND SHORT-EARED OWLS 141

Thinking Like an Owl: GREAT GRAY OWLS WITH GREAT HORNED OWLS 161

Notes from the Field: Insights from an Owl 199
Acknowledgments 201
Owl Resources 203

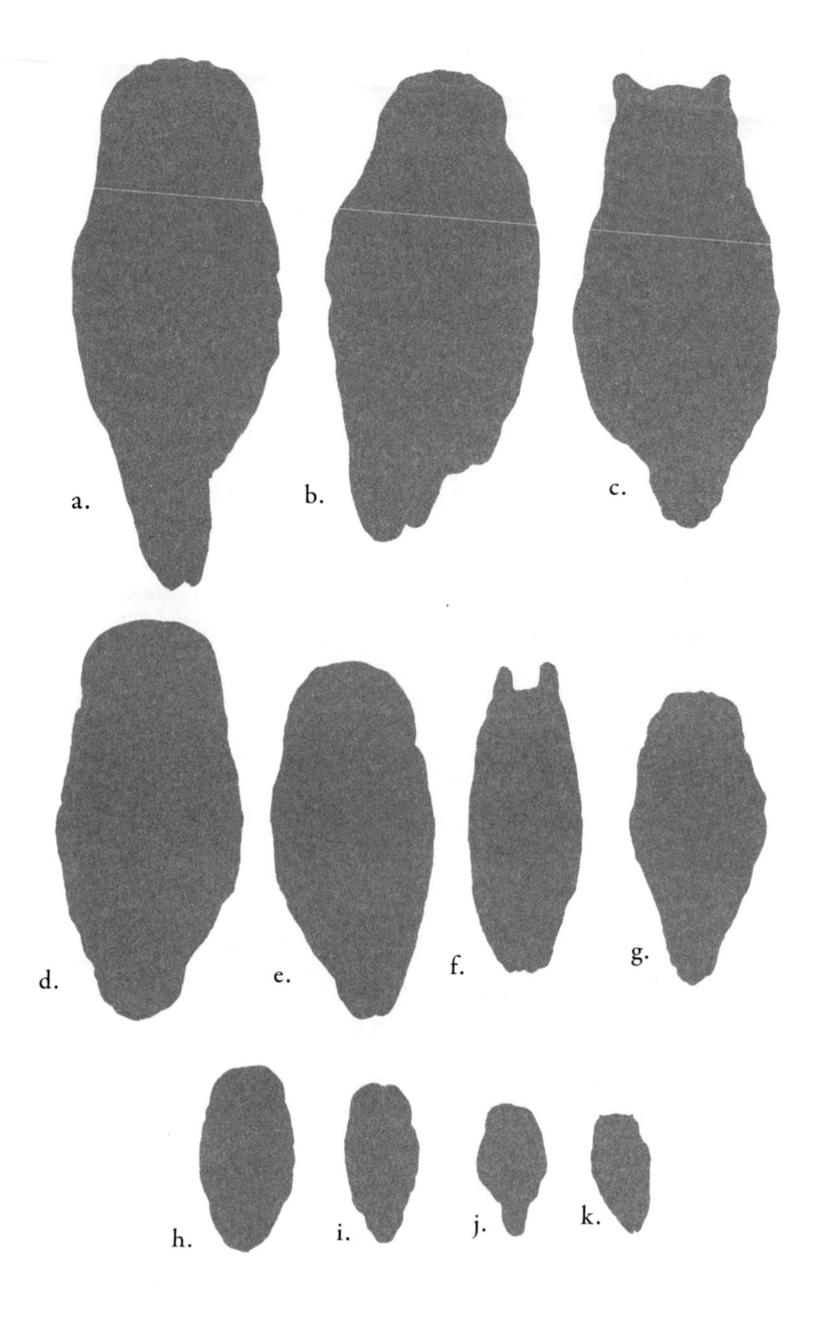
a.
b.
c.
d.
e.
f.
g.
h.
i.
j.
k.

A Comparison of Owls by Size

a. **Great Gray Owl:** length, 24" to 33.1"; wingspan, 53.9" to 60.2"

b. **Snowy Owl:** length, 20.5" to 28"; wingspan, 49.6" to 57"

c. **Great Horned Owl:** length, 18.1" to 24.8"; wingspan, 39.8" to 57.1"

d. **Barred Owl:** length, 16.9" to 19.7"; wingspan, 39" to 43.3"

e. **Northern Spotted Owl:** length, 18.5" to 18.9"; wingspan, 39.8"

f. **Long-eared Owl:** length, 13.8" to 15.7"; wingspan, 35.4" to 39.4"

g. **Short-eared Owl:** length, 13.4" to 16.9"; wingspan, 33.5" to 40.6"

h. **Burrowing Owl:** length, 7.5" to 9.8"; wingspan, 21.7"

i. **Saw-whet Owl:** length, 7.1" to 8.3"; wingspan, 16.5" to 18.9"

j. **Northern Pygmy Owl:** length, 6.3" to 7.1"; wingspan, 15"

k. **Flammulated Owl:** length, 5.9" to 6.7"; wingspan, 16.1"

Introduction

In the dim light of evening, you step out of the forest. A sudden movement startles you, as a large bird on silent wings flies in front of you. Your heart pounds in your chest, not in fear but with the thrill of seeing a wild owl. You walk on, one foot in front of the other, hoping to get a glimpse of the wise bird just ahead on a branch overlooking the trail. To your surprise, the owl remains still, silent, watching. You notice first her round owl face and her curved owl beak, her wings folded by her sides and her long black talons gripping the branch on which she sits. She looks down at you with big yellow eyes. A bit uncomfortable with her deep stare, you wonder what she knows. You sit down on a rock covered with green moss and needles from a nearby fir, to watch and wait, as if for a lesson. You are owling.

"Our association with owls dates back millennia," David H. Johnson, director of the Global Owl Project (GLOW), told me. He has spent twelve years studying owls in myth and culture around the world and has made many interesting discoveries.

In north-central Australia, a unique rocky outcropping overlooks the outback. Here, the Wardaman Tribe believes their creator being, Gordol the owl, first created the world. In southwestern Australia, the Nyungar Tribe protects a standing owl stone, *Boyay Gogomat*, the sacred and powerful creator, healer, and destroyer. In Europe, hunter-gatherer people of the Stone Age carved a Long-eared Owl in the Chauvet Cave, home to the second-oldest cave paintings in France at about 32,400 years old. The owl's long association with the Greek goddess of wisdom, Athena, gave rise to the Burrowing Owl's scientific name, *Athene cunicularia*. For centuries, the Ainu people of northeastern Japan have revered the Blakiston's Fish Owl, the heaviest owls in the world weighing as much as ten pounds, as "the Emperor of the Night" or "the God That Protects the Village." The Mayans wore owl amulets upside down so that the protective owl spirit could look up at the person it was protecting. In Kazakhstan, there exists a mountain range where only female shamans go to connect with the spirit of the owl. The Scandinavian Sami people believe that owls are good luck. And the Native American Navajo believe owl and coyote hold the balance of day and night.

Beliefs, opinions, and superstitions about owls vary widely from culture to culture even today. In South Africa, owls are associated with witchcraft and bad luck; to call someone there an "owl" is the highest insult. In Jamaican folk tradition, if an owl flies by your house it means death. But in Mongolia, owls are powerful spirits that keep away bad luck; owl feathers are tied on baby cribs for protection, and at a popular owl festival each year, women in owl headdresses dance for vitality, prosperity, and wealth. In Turkey, a waitress told David Johnson that if an owl hoots twice around a pregnant woman, it's a boy, if three times, it's a girl, adding, "Doesn't everybody know that?" And when David took a group of fourteen- and fifteen-year-old members of the Umatilla tribe out to see the Burrowing Owls he studies at the Umatilla Chemical Depot in northeastern Oregon, the kids wouldn't even look at the owls at first. In many Native American tribes, the powerful owl is for the tribe's shaman only. Some people even believe there's a connection between owls and extraterrestrials. Writer and artist Mike Clelland, who writes about owls and mythology, believes that there is a synchronistic link between owls and UFO sightings. What is it about owls that makes them so mysterious and fascinating to us?

For more than sixty-seven million years, owls have roamed the earth, flying, hunting, and raising their families in the dark. As the taxonomic order Strigiformes, owls split from the evolutionary branch of the raptors and evolved to not only survive in but thrive in nearly every habitat on the planet, from extreme polar regions to high desert steppe and from deep primeval forests to the farms and neighborhoods associated with human civilization. Owls are divided into two families: Tytonidae, barn owls, the oldest owl species with a heart-shaped face, and Strigidae, typical or true owls, with a round face.

To take advantage of their nighttime niche, these charismatic megafauna of the avian world have developed several unique adaptations common to all owls to aid in living after dark. These same characteristics are what make owls distinct from other birds and make them one of the most attractive species of birds to us humans.

Over the millennia, owls evolved tubular eyes, which face forward and are immovable, and are the reason owls developed the ability to turn their heads 270 degrees. Owl eyes have more black-and-white detecting rods than color cones, allowing them to see in the dark. Their large round yellow eyes, with dark pupils wide enough to let in small amounts of light in darkness, are one of the first things we notice about them. In the human world, large eyes with wide pupils hold a certain attraction both for the viewer and the viewed. Studies show that a person's pupils dilate in the presence of someone they are attracted to. Advertisers dilate the eyes of models in photographs to make their products more attractive by default. Nature, it seems, has prepared us biologically to be attracted to owls by giving them such big eyes.

The face of an owl is another highly specialized feature that holds a unique appeal for us. The facial feathers of an owl form a satellite-like dish capable of funneling sound to the owl's ears, which are asymmetrically placed, one higher than the other, on either side of its head. This adaptation helps owls precisely pinpoint prey in three dimensions, making for a swift, efficient strike. Baby owls often seen bobbing and turning their heads in cute videos on the Internet are actually learning to use their hearing. The human brain is hardwired to recognize faces. We see faces in everything from a bowl of balsamic vinegar and oil to the front of a car. Perhaps in the face of an owl we recognize ourselves.

Another common owl characteristic, zygodactyl talons with a twist—two toes on each foot pointing backward, one forward, and a fourth that can move either way—provides owls with an advantageous structure for forceful grasping and squeezing of prey. With their talons resembling opposable thumbs, owls become efficient, resilient, and adaptable, traits we humans often admire.

The stealthy, silent flight of owls comes from primary flight feathers that have evolved serrated comb-like structures along the leading edge and soft fuzz along the top to break up noisy turbulence. The owl's silent flight has been studied by both the aeronautics industry and the US military. Flight is an adaptation long sought after and admired by humans.

All these specialized adaptations add up to one endlessly fascinating family of birds.

When owls landed in my life, their presence was unexpected. Owls began popping out at me in unusual places. They showed up everywhere, from my Facebook page, where I had never noticed an owl before, to everyday items like mugs, kitchen towels, and candleholders, and a story I was writing about my neighbor's sustainably built house. I had trained myself to pay attention to life's little clues. So I began to learn about these birds that I knew very little about. I attended an owl talk near my home, given at the Bainbridge Island Parks and Recreation Department and learned of two species I had never heard of, Northern Saw-whet Owls and Flammulated Owls. I talked to a well-known birder, George Gerdts, and was surprised to learn that there are Burrowing Owls in Washington and Oregon. I began to read about owls, gleaning information about owl characteristics from individual species accounts. But it went deeper than that for me. As a curious naturalist and nature writer, I wanted to know more about owls than I could learn secondhand. I wanted to experience owls in their natural habitat, hear their calls, and see their movements as they lived their owl lives.

For over a year, I delved into the science of the owl. I worked with biologists to observe and study eleven of the fifteen species of owls found in the Pacific Northwest, traveling to the forests and fields that owls inhabit, and gathering insider knowledge about owl biology and behavior. And I watched their ways, looking into their eyes, searching for some sense of the owl's spirit. What does it mean to be an owl? What do they know? Because they have lived on earth for millions of years, what can they teach us? This is the story of my owl journey.

Neighbors

Northern Saw-whet Owls

The little Saw-whet Owl stares into the dark forest with bright-yellow eyes. She sits, watching from her perch near the trunk of a fragrant western red cedar. Thin white feathers woven with shades of tan and brown lie flat in a pattern outward from each eye, like a stand of trees felled in a windstorm, shaping the iconic disk of the owl's face. She wears a white feathered *V* from bill to brows between her two forward-facing eyes. Her black beak does not protrude from her face like that of an eagle or an osprey, but turns downward in the characteristic right-angled curve of an owl's bill. White flecks amid brown feathers adorn her head while white spots line her brown wings. Her white front is streaked with rust-colored feathers, and her powerful talons, covered in pale-cream feathers, look like kitten paws.

Much about this diminutive owl is surprising. Her feathered facial disk—a hallmark owl characteristic—provides a useful function as a moving radar dish, controlled by tiny muscles in the owl's face, that captures the minute sounds of prey rustling amid the detritus of the forest floor. On either side of her head, ears that look like carnivorous Venus flytraps buried under brown feathers are asymmetrically placed, with the right ear higher than the left—common in many owl species—for gathering sounds funneled from the dish. Because the Saw-whet is a nocturnal species, hunting during the darkest hours of night (not all owls are nocturnal), her lopsided ears allow her to pinpoint her prey both vertically and horizontally, allowing a swift, efficient strike from above.

Her small size, about six to eight inches head to tail, belies her strength. This little owl, weighing less than a roll of pennies, will opportunistically catch and consume deer mice, shrews, the occasional small perching bird, and her favorite, the southern red-backed vole. Once she locates a meal, sometimes heavier than she is, she drops silently from her perch and captures her prey with outstretched talons. She then tears her food into large pieces with her strong beak or swallows it whole. Either way, she eats it headfirst—as all owls do—for the protein contained in the brain. The indigestible bits, like the skeleton, teeth, and fur, are caught in her gizzard and form a "pellet" to be regurgitated about six hours later. It is said that the wisdom of owls comes from their ability to discern what is useful while discarding the rest.

Having moved down from the mountains to a lower elevation for winter, this female Saw-whet on this particular night flies through a forest of tall Douglas firs, western red cedars, and big-leaf maples on the southern end of Bainbridge Island, Washington, directly west of Seattle. Suddenly, she hears an advertising call from a male of her kind. She considers his persistent *tooh, tooh, tooh, tooh, tooh* whistle of an offer. Finding his whistle attractive and knowing it will most likely include a freshly caught vole when she examines the nest cavity he guards, she flies in the direction of the sound to investigate. She darts across an open patch in the trees, coming to rest on the branch of a tall fir.

On the road below we listened, as another owl called in the distance. *Who cooks for you. Who cooks the food?* it asked.

"Barred Owl," whispered our leader, Jamie Acker, pausing only briefly in his whistle to point out the telltale call of the Barred. *Tooh, tooh, tooh, tooh, tooh*, he began mimicking again, the steady, rapid one-tone call of the male Saw-whet Owl advertising for a mate.

We stood in the middle of the road, staring up into the formless sky with fog hanging low over the tops of the Doug firs. All was silent except for the distant moan of a ferry horn from an early morning passenger boat plying the waters of Puget Sound. We tried to stand as motionless as possible. Any movement would rustle the nylon fabric of our clothes, drowning out quiet owl calls. Any loud breath could cover over a slight movement of branches. We all looked east in the same direction as Jamie, listening and watching, five men and me. They were birders from the Washington Ornithological Society (WOS) who'd come to Bainbridge Island, my home for seventeen years, on the 2:00 a.m. ferry to search for owls. "Owling," they called it.

Suddenly, a dark shape with wings silhouetted against the dim, foggy white sky shot across a gap in the trees from one side of the road to the other. We all turned in place with one swift movement, following the small owl. It landed silently in a big fir. Using a spotlight, Jamie briefly searched the trees on the branches closest to the trunks where Saw-whets prefer to roost and then began his rapid call again. I stood staring into the trees, replaying the brief scene in my mind. It was so fast that if I had not been looking into the gap at that precise moment, I would have missed it.

I was used to quick sightings. I had come to owls by the way of whales. Trained as both a scientist and a naturalist to search the vast ocean for any anomaly that would lead to a whale sighting, I had looked for misty spouts, groups of whale-watching boats, and bodies of humpback whales in Hawaii, blue whales off Santa Barbara, gray whales roaming the Pacific coast, sperm whales off the Azores, and orcas swimming in the waters of Puget Sound near my home. It was my desire to help make the waters of the modern world safer for the whales I'd come to know and love that led me to nature writing. Nature writing in turn inspired me to explore and write about other animals and places, like spirit bears and brown bears in the Great Bear Rainforest along the west coast of British Columbia; Bengal tigers in India; polar bears in Churchill, Manitoba; and coyotes on Bainbridge Island.

After I spent some time getting to know some of the finned and four-leggeds, exploring the winged world of owls seemed like a natural next step.

Now, here I stood in the middle of a road during the last moments of dark, searching for one tiny Northern Saw-whet Owl on one of the species' wintering sites around the Pacific Northwest. Jamie Acker shone a spotlight into the trees, and my eyes followed the light, hoping for another look, but I saw nothing. I suspected the owl was hiding, disappointed in the false male's offer that she'd flown all the way across the road to investigate. The seven of us returned to our cars, happy with the brief sighting and full of hope for the rest of our owling adventure.

Earlier that night, I'd waited at a park-and-ride lot for the man known as arguably the best owler in the state; a friend had suggested I contact him. I didn't know what to expect. I'd never done this before. It was cold as I sat in my car, but I didn't wait long before a gray Subaru Forester pulled up beside me. "Are you Jamie?" I asked, just to be sure before I climbed into his car.

"There's no one out here but me," he said.

Jamie Acker is no stranger to driving around Bainbridge Island alone at night. In fact on most nights on our sleepy island, this retired naval submarine officer turned high school physics teacher is the only one on the roads. He purchased a vanity plate for his car with the word "owler" inscribed on it to signal to police and concerned neighbors that owls are all he is after. Once retired from his years of navy service, Jamie was free to turn his eyes skyward and reignite his love of birds, which had started in kindergarten. In 1994, a birder friend, George Gerdts, took him owling for the first time on Bainbridge Island. It was Jamie's first sighting of the tiny Saw-whet Owl that turned his passion into an avocation.

From then on, he has dedicated his free time to the study of owls. "As an engineer, I marvel at their adaptations that allow them to do what they do," Jamie said, explaining his fascination. "From special feathers that channel sound into their ears, to their eyes that see so much better in low light than ours, to feathers on their leading primaries that break up the airflow into small turbulences to reduce noise. Even their talons are arranged to support maximum efficiency during an attack."

Following the required US Geological Survey (USGS) standards for bird banding, Jamie Acker learned how to identify the species, age, and sex of the owls he banded, and documented the number of owls he caught, the hours he worked, and the level of supervision he'd received during the process. He created a research proposal, stating the goals and purpose of his study in fine detail. Then the citizen scientist submitted all of this documentation along with his application to earn a master bird banding permit from the USGS in compliance with the Migratory Bird Treaty Act, under which most owls are protected. He began his study of owls during his spare time, which, due to small children, a wife, and a teaching career, just happened to be the middle of the night. By carefully catching and banding owls, he gathered important data about population, distribution, and territory size for the owls on Bainbridge Island, along with information on the little-understood migratory patterns of Saw-whet Owls. He's spent countless hours wandering the woods in the dark of night, calling and listening for a response.

Jamie and I arrived at IslandWood, a well-known environmental education center where Jamie often captures and bands owls, at about 2:15 a.m. We planned to search for four owl species that night, including the tiny Saw-whet Owls that had changed Jamie's life. I could feel the thrill of excitement running through my body. In my mind, I jumped up and down, clapping my hands like a little girl getting a new bike. On the outside, I tried to maintain my composure.

Once parked, we heard an owl calling: *Who cooks for you. Who cooks the food?*

"That's Gus," said Jamie. "He's eleven."

"How do you know he's eleven?" I asked, accepting without question that the owl was Gus but thinking Jamie knew his age by some quirk in his call that even I could tell was the call of a Barred Owl (*Strix varia*).

"I banded him eleven years ago," Jamie replied.

Jamie called back to Gus, mimicking the owl's sound with his voice. *Who cooks for you. Who cooks the food?* But Gus did not come over as Jamie had expected. "Huh, he must be busy," he said absently as he opened the back of his Subaru, exposing a tangled collection of equipment, including folding chairs and a table, headlamps, a rolling plastic cabinet, mist nets, ropes, a small cage of live mice, an MP3 player with speakers, and other useful odds and ends. "Gus hates nets," Jamie added, to further explain Gus's standoffishness.

I would learn that over the years Jamie had developed a special bond with Gus, who would keep him company during his long hours spent banding owls. He taught Gus to take mice in exchange for not eating the Saw-whet Owls he caught in the nets. Normally, Gus would come at the sound of Jamie's car. But this night was different. For the first time in eleven years, Gus did not come.

About a week later, I saw Jamie at the popular owl talk he and George Gerdts give every year for the Bainbridge Island Parks and Recreation Department. "I found Gus's leg band and some feathers a few days after you were out," Jamie told me, visibly shaken. "He must have gotten caught by a coyote when he went down for a drink." Jamie looked as if he'd lost his best friend, and I felt for him. I understood the special bond that could be formed between a human and a wild creature. And without Gus's cooperation, any Saw-whet Owls Jamie caught for banding could be in danger—a hard fact that Jamie would face in the coming year when the male Barred Owl who replaced Gus in his territory killed a small Saw-whet in the net.

On this night, though, Jamie carried on as if Gus were still around. He disentangled a couple of ropes from the other gear and pulled out three mist nets, folded and secured neatly from top to bottom for easy setup. After he handed me a share of the load, we walked down the dark trail illuminated only by our headlamps to find six eight-foot poles lying alongside the trail for the three long nets. Jamie looped the end of each four-foot-by-fifteen-foot net over the end of a pole, then pushed the pointed end of both poles into the ground so that the net stood upright. When he slid the bottom of the net down the pole into position, the net that would catch the Saw-whet Owls opened easily. He did the same for the other end of the net, while I stood watching, not knowing how best to help.

Once all three nets were set, he turned on the MP3 player and played the loud repetitive male territorial call to lure nearby Saw-whet Owls into the net. Any females looking for a mate would be caught in the net when they flew in to check out the calling male. Males would be caught if they responded to the false male's challenge. In the next forty-five minutes or so, we would know if any of the little owls were listening.

The male Saw-whet Owl uses only his voice to establish his territory and to advertise his whereabouts to any eligible females within earshot. Once he finds a suitable tree cavity for his mate to lay her eggs in—about two weeks before looking for a mate for the season—the male flies around his territory from point to point, calling *tooh, tooh, tooh* at an assertive 160 *tooh*s per minute that can be heard up to about a mile away. Once he feels secure within his established boundaries, he finds a perch near his tree cavity to advertise for females at a slower, more relaxed rate of 112 *tooh*s per minute. When a female flies over to him, essentially agreeing to meet with him, he flies to the nest cavity to show off his home. He then switches to a softer but faster 260 *tooh*s per minute, as if to say *Follow me, follow me.* If she likes him, she will inspect the tree cavity he has chosen. If she doesn't like him, she will refuse his offer and he will call after her as she flies away using his assertive 160-*tooh*-per-minute call.

The Northern Saw-whet Owl is just one of many owl species David Johnson of the Global Owl Project (GLOW) has studied in over thirty years as an ornithologist. Like Jamie Acker, he spent many nights counting the calls of Saw-whet Owls. On one 15° night in Minnesota with fifteen inches of snow on the ground, he sat wrapped in his snowmobile suit on a five-gallon bucket with a boat cushion for a seat. As a small pack of four wolves circled him, so close he could smell the wet-dog scent of wolf just beyond his range of vision, he counted the calls of one little Saw-whet male.

From thirty minutes after sunset to thirty minutes before dawn, David watched, by the light of the moon, female after female fly in to meet this male. He listened carefully as the male switched to his softer, faster call, audible only within about one hundred feet. He was astonished when, one by one, each of the five females refused him. Curious as to why the male had been rejected over and over, he climbed the aspen tree to the Pileated Woodpecker hole that was the nest cavity the male defended. Inside, he found there was an old honeycomb partially blocking the entrance. The females could not get in or out.

"For some males, the most attractive males with the best nest sites, they get mates sooner and stop calling," David later explained to me. "For this male, this was all he had. He just kept calling until he was hoarse."

Jamie and I walked back to the car, where he had set up his Saw-whet banding station. He retrieved the folding table and chairs from the trunk and arranged an array of equipment on the table—a catalog filled with pages of individual owl statistics, a pen, a digital caliper, a food scale, a small blue bag, and an empty frozen-orange-juice can well worn around the edges from years of use. I tried to imagine the use for the orange-juice can.

"I need to meet the others at the gate. Do you want to come or stay?" he asked. The "others" were the five men from WOS joining us for this evening's owling field trip.

"I'll wait here," I said, slightly surprised by my response. I'd had a fear of the dark woods since I'd accidentally seen the trailer for *The Blair Witch Project* years before. I'd been jumpy all week just knowing that while owling I'd be in the dark for hours. But I was determined not to embarrass myself. Now, as I sat in the forest at the makeshift field station, waiting alone with only the light from my headlamp to keep me company, I was not afraid, listening to the recorded call of the male Saw-whet Owl. I was relieved, though, when I heard voices and turned to see my fellow owlers-for-the-evening moving my way.

We introduced ourselves and shook hands. There was Bill, a photographer; John, an engineer, who was studying biology for the fun of it; and three other WOS owlers, Randy, Matt, and Ken. I knew that not one of these men would I ever be able to recognize at a cocktail party. Meeting them in the dark created a perplexing dynamic of faceless voices speaking without the listener knowing who said what. Except for me. Anytime I opened my mouth, anytime I asked a silly question, everyone would know who had spoken. The good thing was, they would never be able to recognize me either.

Once enough time had passed to allow the Saw-whet Owls ample opportunity to investigate the recorded call, we walked single file along the trail to check the nets. I listened closely, but I heard nothing but the recording as we hurried down the trail. When the first net came into view, my heart skipped at the possibility of my first owl being so close. My heart sank just as fast when I saw the empty net. But, of course, there was another net, and my heart raced as we ducked around a pole to check the second net. It, too, was empty. Would we catch any owls? Was anyone listening to our call in the dark? I still had hope.

I was second to last in the single-file line that led to the third net. I followed the men closely as they made their way toward it. Then they stopped and formed a semicircle. To get a better look, I walked around to the other end. And there I saw a tiny owl hanging upside down like a bat, wrapped helplessly in the fine mesh. A quick breath caught in my throat.

"This one really did a number on herself," said Jamie as he carefully disentangled the owl, so as not to tear its feathers. As Jamie held the owl in one hand, it clacked its beak in a show of force larger than its ability at the moment. I breathed, then smiled at the little owl's fierceness. I saw its face and its bright-yellow eyes wide open, staring straight at me. I'd never seen such a small owl. I never even knew such a small owl existed. Yet in that moment, looking into those big yellow eyes, I felt a connection to this being.

As we walked back toward the banding station, I wanted to maintain my scientific perspective, but I was losing the battle with my heart. "It's so cute!" whispered a male voice beside me. I smiled at the unexpected sentiment from the ornithologist that mirrored my own thoughts. I turned to see who had said it but in the dark, I couldn't tell.

Saw-whets are named for their rasping call, reminiscent of the sharpening or "whetting" of a saw with a file—a name that I suspect held more meaning for the New World settlers who first described it than it does for modern-day readers. But this small owl remained silent as Jamie carefully carried it back up the trail one hundred yards to the banding station. He held the bird securely in his left hand, with the owl's legs lightly between his pinkie and ring finger. With his thumb around the owl's shoulders, his hand covered its right wing while his fingers secured the left. First, he read the numbers off the owl's tiny leg band, a fitting around the owl's leg with a diameter of less than a quarter inch, the thickness of three US nickels stacked one on top of the other. "One-oh-one-four dash four-three-five-four-eight," he said out loud as he flipped through his thick three-ring binder with pages of owl statistics from each bird he'd ever caught and banded. "I've caught her before," he said, reading his notes. "A little more than a year ago . . . I don't know if she stayed here all year or if she came back again, but it's significant that we caught her again."

I did not know it then, but this seemingly small bit of information added with other bits from a simple band around a bird's leg could over time reveal so much about where it has been, how often it returns to the same location, and how long it lives.

Bird banding, the practice of placing a numbered silver band around a bird's leg, began somewhat accidentally in 1595 when one of King Henry IV's Peregrine Falcons, which had a metal ring around its leg, went missing while hunting in France. The falcon was found the next day in Malta, some thirteen hundred miles away. To get that far in twenty-four hours, the Peregrine Falcon would have had to average fifty-four miles per hour nonstop. After that discovery, "ringing," as bird banding is called in Europe, then became a popular curiosity of the time.

In 1728, a Grey Heron that had been banded in 1669 was caught, revealing that a Grey Heron, the European relative of the North American Great Blue Heron, could live nearly sixty years. Several years earlier, another Grey Heron with a band on its leg from Turkey was captured by a falconer in Germany. At some point in its life, the bird had made the twelve-hundred-mile migration from Turkey to Germany. Through the use of the leg band, the lives of our avian neighbors were being revealed to us one puzzle piece at a time.

It was John James Audubon, famous wildlife artist and namesake of the National Audubon Society, who first banded birds in North America in 1803. By tying a silver cord around the legs of chicks in a nest of Eastern Phoebes, a small brown-and-white flycatcher, he was able to individually identify the Phoebes each year when they returned to his Philadelphia neighborhood. In 1902, Paul Bartsch was the first to label the bands he placed on one hundred Black-crowned Night Herons with the instruction "Return to Smithsonian Institution," thus creating a system for collecting information from banded birds.

Banding birds also helped these early ornithologists recognize that they were seeing the same birds even though the birds' plumage looked different at times due to season or age. For example, juvenile Saw-whet Owls, with their solid chocolate-brown heads and rust-colored chests, can be easily mistaken for another species. For the first six months of their lives, only the white *V* between bill and brows gives any clue as to their parentage. Once a bird is banded, it is easily identifiable for years to come when recaptured.

In 1920, Frederick Lincoln began organizing the first North American bird banding program as a result of the Migratory Bird Treaty Act, signed in 1918 by both the United States and Canada. It remains today as a cornerstone of ornithological research, conservation, and management. Utilizing this vastly expanded network of bird banders and their collective information, Lincoln was able to propose the "Lincoln Index," a method of estimating the numbers of birds in various populations. He also defined the idea of North American flyways for migrating birds.

In the twenty-first century, technological advances—in radio telemetry, tracking the location of birds with radio signals; geolocators, measuring sunlight; and solar-powered mobile tracking systems, like GPS and GMS, using satellites and cell towers—have led to the development of tags and transmitters that are now small and light enough to place on the smallest of owls. Scientists now have unprecedented access to owls and other birds through minute-by-minute transmissions, revealing some amazing behavior never before witnessed. Today, with the use of these new technologies, we are poised on the edge of the next big breakthroughs in avian biology, just like the curious renaissance birders once were when they made their first big discoveries with the help of slim silver leg bands.

After years of using the simple but proven banding technique on Saw-whet Owls, biologists are beginning to get a clearer picture of the owls' habits and behaviors. Northern Saw-whets Owls, *Aegolius acadicus*, were named for the colony of Acadia, now called Nova Scotia, where the owls were first described. Scientists have discovered that, in fact, they are found across the continent, from Mexico to the southern edge of Canada's boreal forests. We also now know that in the Pacific Northwest, some Saw-whets seem to move seasonally north to south and from high to low elevations, from the Cascade Range and the Olympic Mountains in the spring and summer to sea-level islands like Bainbridge in the fall and winter. Others may wander nomadically. One little Saw-whet banded in Montana ended up in Massachusetts. Another group, a subspecies of the Northern Saw-whet, *Aegolius acadicus brooksi*, lives year-round in Haida Gwaii, formerly known as the Queen Charlotte Islands, off British Columbia. Saw-whet Owls seem to prefer mature or old-growth mixed evergreen forests and deciduous woodlands for the plentiful cavity-nesting prospects there with hunting opportunities along forest edges near rivers and streams.

Through careful observation over the years, ornithologists have found that male Saw-whets establish and defend their territories beginning in March. Because owls do not build their own nests, but use a nest or a cavity left behind by another bird, Northern Saw-whet Owls in the Pacific Northwest choose territories that usually include at least one viable nesting option, possibly more, in a tree cavity made by flickers or Pileated Woodpeckers.

Once a pair of Saw-whets bonds, the female will begin laying her eggs, one every one to two days until four to seven eggs are laid. Unlike some birds, who postpone incubation until all the eggs in the clutch are laid, owls and other birds like hawks kingfishers, egrets, and herons will begin to incubate their eggs immediately after the first egg is laid, causing chicks within the same nest to be up to two weeks apart in age and at varied stages of growth. Known as "asynchronous hatching," this strategy ensures that at least some of the chicks from the clutch will survive even in hard years when there is only enough food for a few and the natural competition of survival of the fittest kicks in. In years of plenty, even the youngest chicks will survive.

With the female owl sitting on the nest continuously for the entire twenty-six- to twenty-nine-day incubation period, the male dutifully brings food to his nest-bound mate in a family strategy reminiscent of one of our own human approaches where the father works while the mother cares for the young. Once the chicks are hatched and able to generate their own body heat, or thermoregulate, at about eighteen days old, the mother Saw-whet Owl moves from the nest cavity but remains nearby to help with the feeding. At thirty-three days, the chicks have grown flight feathers and are ready to fledge. The father continues to feed the chicks for another month or so while the mother leaves the family in order to recover her strength or, in years of plenty, to raise a second brood.

All of these owl ways were still a mystery to me the night I stood at the banding station in the dark forest on Bainbridge Island, surrounded by my fellow owlers. I watched, fascinated, while Jamie took the next critical measurements. "I know from the first time I banded her that she's a female. I will take all these measurements again to confirm what I thought originally," he explained as he gathered the necessary equipment. "The female is larger than the male, so I will measure her wing chord and take her weight to be sure I'm right."

Of all the facts about owls that I would learn, one of the more interesting to me was the fact that while for most other birds the male is larger than the female, for most owl species, including the Saw-whet Owl, the female is larger than the male. There are several theories to explain this evolutionary trait, known as reversed sexual size dimorphism. One is that the size difference may reduce competition for prey between males and females. As the smaller of the two, the male may be a faster, more agile hunter, a trait that would serve him well in providing food for his hungry mate and chicks. Another theory is that the female may need to be larger to carry more weight in preparation for her long period of incubation. Lastly, a larger female may be better able to defend her nest from would-be predators.

Jamie went to work, carefully measuring the owl's wing using a universal wing ruler, while the rest of us watched in silence. He held one of the owl's wings out slightly from its side, still naturally folded at the bend, or wrist. He then placed the lip of the ruler under the wing at the last joint in the wrist, just above the primary feathers. Jamie held the bird securely yet gently, letting the wing rest on the ruler as he measured from this joint to the tip of the longest primary feather. Every now and then, the owl would voice its frustration at being held captive, registering its dissatisfaction with a sharp clap of the beak. *Clack . . . clack!*

A female Saw-whet Owl's wing is longer than 143 millimeters or about five and a half inches, while a male's wing is slightly shorter than five inches at 128 millimeters. (Of course, there is always some overlap on the bell curve between the sexes.) The length of this little owl's wing indicated that it was likely a female.

One more measurement would confirm that Jamie had determined the owl's sex correctly. While holding the little owl in one hand, he reached with the other hand for the scale and the orange-juice can. He set the can on the scale and took its weight. Then, so that she couldn't fly away, he slid the owl headfirst into the can, tail in the air to keep the bird secure, and took that weight on the scale. The weight of the empty can would be subtracted from the overall weight of the owl plus the can to calculate the owl's true weight. I never would have guessed that this was the use for the empty frozen-orange-juice can!

"Ninety-three grams," Jamie said proudly as he extracted the owl from the makeshift cage. Now we knew for sure it was a female. The average weight for a male Saw-whet is closer to seventy-five grams. Next, he turned the little female on her back to examine the pocket of fat under her wing and over her breastbone. "She's a healthy bird," he announced. Jamie also took measurements of the Saw-whet's tail with the universal wing ruler, then used a digital caliper to measure her black beak.

Then he was ready to assess the age of the female, using a technique his studies on Bainbridge Island had helped him pioneer. In 1982, ornithologist Bruce Colvin had the idea to estimate a Saw-whet Owl's age by using ultraviolet light to fluoresce the underside of the primary and secondary feathers on an owl's wing. He knew that these feathers contained a group of pigments called "porphyrins" that degrade from prolonged sun exposure. Porphyrins are most abundant in new feathers and will glow when held under a black light.

Owls lose only a few primary and secondary feathers each year, so they are never rendered flightless, as some birds are during their molt. Some owls of different ages have different molt patterns that will show as a fluorescent pink from the blood that remains in each newly formed feather or a chalky white in older feathers. For years, Colvin's findings remained unpublished and were used only by those who personally knew of his technique. Jamie Acker was one of those researchers. He used the ultraviolet aging technique on Northern Saw-whets, Barred Owls, and Northern Pygmy Owls with success.

Now Jamie used the ultraviolet light on the underside of the little Saw-whet's wings. She remained calm as he turned her over again and unfolded her right wing. "You see how it fluoresces pink?" he asked as he ran his finger across her outstretched wing. "That's the pattern of an 'after second year' bird," he said, pointing to the new primary and secondary feathers.

Once he'd completed all his measurements, Jamie placed the bird in a dark-blue bag and hung the bag on the side of the plastic cabinet. "She'll be fine there while I check the nets. Her eyes need time to readjust to the dark. We don't want to send her off into the forest while she's night-blind," said Jamie as he walked into the dark again.

"Wow!" said one of the men as we stood in a small circle around the table, talking quietly so that we didn't disturb the owl hanging in the bag nearby. "That's the first time I've ever seen quantitative measurements taken on a bird. I usually just keep track of the species I see."

"Yes, very interesting," said another.

"What did he say about the yellow eyes?" I asked, trying to gather as much information as I could.

"There's a theory that the yellow color deepens as they age. He measured them against a chart he had," another voice explained.

"And that bird was 'an after second year'?" I asked the circle of men. "What does that mean?"

"It means she's older than two years but not yet three."

Geez, that's confusing, I thought. *Why don't they just say she's one, two, or three?*

A few minutes later, Jamie arrived with another Saw-whet Owl in hand. He repeated all the measures he'd taken on the first owl, checking wings, tail, beak, and weight, and fluorescing the new owl's primary and secondary feathers. This owl also had a band, and again all of Jamie's measurements agreed with his assessment from the first time he'd caught the owl. It was another healthy female adult, but slightly younger, a "second year." He placed the owl in another bag and hung it beside the first one, then went off to take down the nets. After he'd disassembled the banding station and loaded his gear back into his car, he took the two little Saw-whets into the forest to release them back into the wild, each having made her contribution to science.

At 5:00 a.m., after a thrilling start to our owling adventure, we drove to various locations around Bainbridge Island, searching for Saw-whet, Barred, Barn, and Great Horned Owls. On the road, we stood in silence, listening for owls. In the distance, I could hear the low moan of ferryboats on early morning runs around Puget Sound, restless chickens in their coops, and dogs barking at the night. Every time we got back into the car, I'd thought of more things to ask Jamie. During each ride, I pestered this quiet man with my excited questions.

Two more Saw-whets responded to Jamie's calls, flying in to check out the new male on the block. We met Laverne, another of Jamie's banded Barred Owls, like Gus. And we heard a wild Great Horned Owl responding to a captive Great Horned, forever injured and living as an educational bird.

As I stood in the dark listening, I sometimes heard owls but other times missed their subtle talk. I tried to carry the new owl facts in my mind, holding them tight as if I might drop the many precious gifts I had. As the night turned to dim dawn, I felt the mist of the heavy fog on my face and hair on this cold January night.

Somewhere out there in the forests of Bainbridge Island were Saw-whet Owls living their owl lives. Some would move on shortly, maybe up the nearby mountains or north into British Columbia. Some would stay on the island where I had chosen to live. They would hunt for red-backed voles. They would search for perfectly sized tree cavities to call home. They would call for and find mates and raise their young. All in the dark of night. They live alongside us here in the Pacific Northwest—neighbors we would be fortunate to meet.

Flames of the Forest

Flammulated Owls

Darkness closed in around me as I made myself comfortable. I leaned against my backpack at the base of a fir tree in the Naches Ranger District of the Okanogan-Wenatchee National Forest, about thirty miles northwest of Yakima and about twenty-five miles east of Mount Rainier. It was now just past sundown and, at an elevation of about 2,700 feet, growing cold fast on this late spring night. I had put on all my layers at once, including hat and gloves, but already my teeth were chattering before I settled in for what I knew could be a long wait.

Somewhere in the distance I could hear the rhythmic call of the tiny male Flammulated Owl, echoing through the foothills, the deep hollow woodwind-like *hoop* every fourth beat reminiscent of Pan blowing his pipes. Whether it was a real owl calling or a recorded playback, I couldn't be sure.

My companions—Dave Oleyar, a thirtysomething PhD from the University of Washington, now working with HawkWatch International, and summer volunteer intern Annabelle Bernabe from GLOW—sat under other trees somewhere nearby but out of view. The three of us sat under the stars on this clear night, hoping to recapture one of the thirteen male Flammulated Owls that Dave had fitted over the last two nesting seasons with a tiny geolocator "backpack" that looks more like an inch-long blue antennae than a backpack. The geolocator is carefully attached to the owl's back to record changes in the amount of daylight as a means of tracking the tiny owl's travels. Where these Flammulated Owls spent their winter months was the million-dollar question that Dave Oleyar hoped to answer.

I stayed as still as possible, trying not to move so I didn't make any sounds that might frighten the owls or, worse, give away the trap we hoped to lure them into. The trap consisted of three twenty-foot-high mist nets, two sets of speakers and MP3 players, and one Styrofoam owl covered with feathers, which Dave had glued together and nicknamed Flameo.

Flammulated Owls are territorial in nature and spend a considerable amount of time and energy defending the area around the tree cavities they have chosen. At this time of year, their intentions are twofold—one, to defend their chosen territory, and two, to attract a mate. Another owl advertising near their territory, including fake owl Flameo, would be worthy of investigation. Dave was counting on this owl reasoning when he devised our current strategy. He had spent many years studying Flammulated Owls.

For his master's work, Dave had studied the impact of forest clearing for ski area development on these tiny owls during the 2002 Winter Olympics preparations in Ogden, Utah.

Flammulated Owls are highly nocturnal, more so than some other owls that hunt during the crepuscular hours during dawn and dusk. So earlier in the evening, just before dark, we had set mist nets at angles perpendicular to the directions that Dave surmised the owls would come from when flying in to check out the audacious Flameo, who dared intrude into their space. Having watched as Jamie Acker set mist nets while catching Saw-whet Owls on Bainbridge Island, I had some idea of what to expect. But these nets were twice as high, adding complexity on top of trickiness to get them to stand upright.

First, we joined two sections of ten-foot electrical conduit pole that Dave had purchased from a hardware store. Then we slid the loops from one end of the mist net onto one pole and placed that pole over a rebar stake that someone had hammered into the ground. One person held that end up while someone else repeated the process on the other end. We then attached guylines to the poles to steady them against the wind—and the owls that we hoped would come flying in. Finally, we wired Flameo to a branch just above two speakers and an MP3 player. When the sound was turned on, a tiny mad Flam would come swooping in, only to be trapped in the net. That was the plan anyway. For all I knew, it was a good one.

The cold crept farther through my layers as the wind picked up, and I shivered both for warmth and to keep myself awake.

During the four nights I worked with the team, we began each day at sunset and worked until about two or three in the morning, a schedule I had clearly not yet adjusted to keeping. Periodically, Dave would rise from his seat across the small clearing, turn on his headlamp, and check the nets. This was a good time to stand, stretch, and warm myself a bit more by moving.

"Did you hear the owl?" Dave asked during one break.

"No," I said. "I couldn't tell the difference between the recording and the real owl."

"I think he was sitting somewhere above you in the tree," said Dave. "They actually sound quieter and farther away when they're coming in to investigate. They fool you into thinking they're far away so they can sneak up on the intruder," he explained about this tiny ventriloquistic owl. I found all this owl logic fascinating, and it was beginning to make sense. Slowly the owl puzzle pieces were starting to fit together.

Flammulated Owls are the second-smallest owls in North America and the smallest in the Pacific Northwest. At about six to seven inches long, they weigh nearly fifty grams, not quite as much as a half a roll of pennies. These tiny owls range in color from reddish to gray, depending on the habitat they live in. For those that live primarily among ponderosa pines in the Southwest, the Flammulated Owls' reddish phase—or flame color, which some claim gives rise to their name—camouflages them perfectly. In the Pacific Northwest, among the tall Douglas firs in the eastern foothills of Mount Rainier where the owls have adopted a gray disguise, some people say the name comes from the flame-colored V-shaped stripes on their backs.

With big voices for such small creatures, the Flammulated Owls' calls can be heard for over half a mile, echoing through the forest. Black eyes that seem just a bit too large for their squarish heads peer into the dark of night. With a keen sense of hearing, they hunt among the branches and needles of pines and firs for moths and search the forest floor for crickets and grasshoppers, scorpions, and beetles. They will also catch nocturnal insects in midair, flycatcher-style. The odd mouse has found its way into a nest or two, but they're known as the only owls to forage primarily on insects.

We waited and watched, sitting under our respective trees, checking the net every now and again, only to come up empty in the end. We took down the nets and gathered our gear for the half-mile hike back to the Jeep in the dark. We would try again tomorrow. For tonight, I was just thrilled to have heard the call of the Flammulated Owl, even if I wasn't sure it was real.

On the way back to the US Forest Service cabin where we were staying, Dave explained a bit more about Flammulated Owls. "They're just coming back into this territory from somewhere in the Mexican mountains. So we might be just a little early," he said. The females would begin arriving a few weeks later after the males had established their territories. "The nesting and hatching of chicks is all well synchronized with their prey, mainly moths and other insects, hatching. That's why global climate change could be a problem. It causes a mismatch between prey and predator. Owls need an abundance of food when their eggs hatch."

I thought about the tiny Flammulated Owls who time their migration by length of day, gambling their lives and the lives of their families on the "Goldilocks zone" of their prey—the time that insects reproduce when there's just the right temperature and amount of moisture.

The next day, as I sat on the lawn enjoying the sunshine among tall firs in this mountain forest, I watched as swallows dove, catching tiny bugs, and I felt hopeful that there might also be moths or other nocturnal insects around.

I had started to notice the activity of all creatures and the sky in a way I never had before. Each swoop of a swallow and buzz of a bug took on new meaning. Tall storm clouds and the rumble of thunder didn't just mean we wouldn't trap owls that day but also had some larger significance for the health of the owls themselves. If it was too wet or too cold, there would be an effect on the insects they depended on. Was the effect helpful or harmful? I couldn't say. Yet I was becoming conscious of how all of life worked in harmony, in a synchronicity of species for the good of each. I was learning how all of this fit together for the Flammulated Owls I was now enchanted by.

I found myself hoping the storm would either blow through or hold off. Netting owls in wet weather could cause damage to fine feathers, allowing rain to soak through and cause cold and discomfort for these small owls. Fortunately, as the day progressed, the showers dissipated enough that we decided to give trapping a try.

We would be joined on this night by Markus Mika, who had done his PhD on the genetics of Flammulated Owls on their wintering grounds in Mexico and their breeding grounds in Nevada and Utah, using a collapsible net that he could easily maneuver on his own. He discovered a unique haplotype, an inherited group of genes, in an isolated mountain range in Mexico, which could in the distant future be a marker for a new genetic split, creating another species of this tiny owl.

Currently, the owl world is awash in a debate between scientists who believe genetics should determine what species an owl belongs to and those who believe physical characteristics like an owl's call should determine species placement. Today, according to GLOW, there are 225 owl species with 417 subspecies worldwide, with more being discovered and reclassified every year after much scientific debate. Due to technological developments that have allowed scientists to study DNA, the Flammulated Owl was recently moved from its former placement, with Screech Owls, in the genus *Otus* to its own genus, *Psiloscops*. Because of genetic differences from the Screech Owl, along with differences in call and other physical characteristics, the tiny Flammulated Owl is now known scientifically as *Psiloscops flammeolus*, a rare agreement in the owl species discussion.

Tonight, Markus and Annabelle would work in an area where Markus had heard some calls earlier. Dave and I, joined by two helpful Forest Service "hooters" taking a night off from calling for Spotted Owls, would trap where Dave had tagged an owl the year before. Flammulated Owls seem to show high site fidelity, meaning that they return year after year to the same territories, making the odds of recapturing an owl with a geolocator all the more probable.

We hiked up a steep hill and set the nets as we had done the previous night. We each found a seat where we could "see" the nets and settled in for a long wait.

The bugs were out in force tonight, buzzing and swarming in the light drizzle that had moved in, and attracted the attention of a small mouse-eared bat that got caught in the net. When Dave shone the light on the net, it was so small I thought it was a moth. I had always loved bats for their echolocation ability and was excited to see one up close, especially since owls were scarce.

Dave quickly disentangled the bat before it could do too much damage to the net. Holes in nets are a problem if they grow too big. After releasing the tiny visitor, we returned to our places, only to recapture the same bat a few minutes later. This time I saw it hit the net, and it was soon disentangled for a second time. At least now I had some idea of what to expect should an owl ever hit the net.

Then the rain started in earnest, coming down in larger, heavier drops. In the distance lightning flashed. We wouldn't catch any owls on this night. We quickly took down the nets and headed down the hill. Markus and Annabelle were still somewhere out in the forest, and we had to pick them up before we could leave. The hooters drove us up the valley to where they had parked the Jeep. Dave and I unloaded the nets and waited for Markus and Annabelle to return.

By this time, it was cloudy but dry as we stood in the dark, staring into the night sky. "Did you hear that?" he asked.

"No," I replied.

"It's a bit far, but it's a Flam."

I held my breath as I listened deep into the night.

Hoop . . . hoop . . . hoop, called the Flam. Then a more irritated *Hoo-hoo Hoop. Hoo-hoo Hoop*, as if now his territory was clearly being violated.

"I hear it," I said, excited to have Dave there to confirm that it was the Flammulated's distinctive call echoing through the valley. "What a big voice," I whispered to Dave, who smiled and nodded.

I leaned against the truck, looking up into the now-starry sky. A chill settled in my core, yet I could feel a warming within my heart. Here I felt alive in the darkness in a way I had not for some time.

As a single mother raising a daughter, I'd fallen into wrestling with the daily struggle to get through the day rather than really embracing my life. Some days it was difficult just to get out the door in the morning to get to school or a meeting on time. But here I could remember myself again, come back to the life I'd once seen for myself—out in the wilderness, exploring the world and all its creatures. I stayed with this new feeling for another moment, trying to memorize this space so that I could find my way here again. The call of this little owl had stirred my wild soul.

Hoo-hoo Hoop . . . hoop . . . hoop, called the Flammulated, sounding louder as it flew toward us.

"Do you hear that?" said Markus, walking out of the forest with Annabelle, just down the road from where we waited, as the irritated Flammulated male continued his call to defend his territory. "I found him up there, and he followed us down the valley. Never did catch him," he said, explaining how the owl had been lured in by the fake call but had somehow avoided the net. Apparently, netting a Flammulated Owl was more difficult than I suspected, even if you could set a net right beside one.

The next night we tried again, this time in a third spot. During the off-season, the hooters had placed nest boxes in some tall Doug firs. The boxes were modeled after a nest box that Dave had successfully placed in northern Utah, where he now lives and works. The hooters joined us for another night of Flam-catching, to help us find the nest boxes.

We found that none were in use, but it was still early in the season. Dave also checked a few nest cavities in nearby trees that looked to be likely homes. Picking up a large stick from the forest floor, he knocked on one tree with a nest cavity. "Flam hammer," he said. "Flams will stick their heads out to see what's going on, just like Saw-whets." No Flams emerged.

Finally, Dave found a beautiful nest cavity in an old, bare snag standing alone in a forest opening on a steep slope. We placed two of the nets parallel to one another and one perpendicular to the other tree. Flameo sat in the center, calling his fake Flammulated call, while we found our places under surrounding trees.

Just before sundown, we heard the call of a live Flammulated Owl in response to our false owl call. It was not nearby, but it was a start. I listened closely, as other calls sounded around the forest. At one point, we had as many as four Flams calling at various distances from the nets. As I listened to the *hoops* of Flammulated Owls, I stared at the stars. It was not as cold as it had been previous nights, as the days were warming on the east side of the Cascades. The International Space Station drifted overhead, and I watched its bright path. I thought about the Flams carrying the tiny tags as they flew over three thousand miles from Mexico. Did these Washington Flams fly up the western slopes of the major mountain chains—the Rockies, Sierras, and Cascades—like the other Flammulated Owls from California? Had they joined Utah and Nevada Flams in the Central and Southern Mexican states of Guerrero, Mexico, Oaxaca, Hidalgo, and Veracruz for the winter? Finding these tagged owls from Washington meant a deeper understanding of the Flammulated Owl's life. No tags meant no data.

At midnight we still had not caught any owls, and on this night Dave decided to take a walk to see if we could get any closer to the Flam calling about a quarter of a mile away. Tonight we would try Markus's method of finding an owl and setting a net.

I tried to stay close to the group as we walked into the forest. In the dark, it was difficult to get my bearings.

"Which way would you go to get back to the nets?" Dave asked me at one point in our cross-country journey.

"That way?" I said, taking my best guess and pointing in the opposite direction from where we'd come.

"You stay with me," said Dave.

We hiked for about forty-five minutes toward an uncooperative Flam before we turned back. "We'll come back here tomorrow night," said Dave as we gathered our gear. "This is an active spot." It was still dark as we headed down the hill, listening to the passionate calls of the tiny owls.

Three nights and no owls. But we remained hopeful. There were many owls in the forest; it was just a matter of outthinking them. For now, though, we fell back on the time-honored tradition of superstition. The tradition for new volunteers was to eat an orange marshmallow Circus Peanut, as Annabelle told me she had done upon her arrival. I had not been properly initiated because all the Circus Peanuts had been consumed. So I had to swear to eat one if we could find another bag in some store out here in the wilderness.

I had learned over the years, working with people studying many different species, that the minds of biologists worked in mysterious ways. If luck was what we needed to catch a Flam, then I was willing to do what I could to conjure it up. Like a baseball player wearing his lucky hat, who were we to test the fates? Much to my displeasure, we found a bag of Circus Peanuts the next day at a small restaurant near our cabin. As promised, I upheld my end of the bargain by playing along and eating one of the overly sweet orange treats.

On my last night with Dave, Markus, and Annabelle, we hiked to the spot we ended at the night before, up the steep slope to the nest cavity. I wasn't sure I believed in the power of Circus Peanuts, yet I felt confident that I would see my first Flammulated Owl. We set up the nets as we had before, with Flameo in prime position, just before darkness set in. Once again, we found our respective seats, where we could each have a clear view of the nets against the clear night sky. Dave turned on the recorded call. I looked up at the darkening orange sunset. Then wham! Something heavy hit the net. I saw the tall, flimsy conduit poles swaying with the impact. I saw Markus run to the net first with Dave following close behind.

"What was that?" I asked.

"It's an owl," said Dave. "He flew in right over my head. I heard him in the branches." He took down one end of the net to make it easier for Markus to disentangle the owl. I headed downslope to see the owl Markus held in his hands.

"He's so tiny!" I whispered in disbelief. I had thought Flameo had been made to scale. But he was more the size of a Saw-whet Owl. This tiny Flam made the Saw-whet I had seen seem enormous. And he was so cute, with his large black eyes staring at us, as if his fate was sealed. "This little owl made all those loud calls?" I said.

"Yep. They're the pinnacle of evolution," Markus joked. Even though the owl didn't have one of Dave's backpacks on, we were all encouraged by this capture. We moved away from the net for Dave to process the owl.

First, he weighed the little bird, which was a healthy fifty grams, confirming it was a male. Like all owls, female Flams are slightly larger than the males, at sixty-two to sixty-five grams. Then he measured its wing (5 inches) and tail (2.3 inches), as I had seen Jamie Acker do with the Saw-whet Owls. Finally, he put a band on its tiny feathered leg.

Dave handed me the tiny owl. It fit in one hand. "Leigh gets to let this one go," he said. "Hold its legs lightly between your fingers like this." He showed me how to hold the owl between my middle and ring fingers, making a fist to hold its legs secure. As I positioned my other hand around the Flam, I admired his intricately camouflaged feathers, the color of gray bark, and the ruddy stripe on his back. His oversized dark eyes stared up at me while Markus took our picture.

"Take him into the forest a bit away from the nets," Dave said. "Hold him in your hands. He'll sit for a bit and adjust before he flies." Dave explained that the owl was in a light stupor but that he would recover.

My heart beat hard as I walked into the forest by myself, trying to take in the moment with this tiny wild thing. I wanted to say something meaningful to the owl to let him know I meant no harm, but all I could say was, "You're so beautiful."

I breathed deep and tried not to move too much as I opened my hand. I so wanted to provide a safe space as the tiny owl recovered. He sat in my hands for what seemed like a long while but was probably no more than a minute or two. Then he spread his tiny wings and flew off into the dark night. *Good luck*, I thought. *May you find a beautiful owl wife and have many owl babies.*

Then a chill ran down my spine as *Who cooks for you. Who cooks the food?*, the telltale guttural call of a Barred Owl, rang through the forest. I hoped the tiny Flam had found a safe place within the branches of a fir. Barred Owls, recent arrivals to the Pacific Northwest, as well as Northern Goshawks and Cooper's Hawks, are known to prey on the smaller Flammulated Owls.

"He flew," I said with a smile as I returned to the team. "Did you hear the Barred?"

"Yeah, we heard it. Last year I heard a female Spotted being chased by a Barred. That's the sound of extinction, I thought to myself," Dave said before hesitating to ponder the Barred's presence. "I don't want to take a chance catching any Flams here. Let's take the nets down. We'll finish early tonight. Let's go celebrate." We took the nets down and packed our gear.

The team would not recover any of the thirteen backpacks during their two weeks of trapping, yet Dave remained hopeful that someday they would. So for now the winter whereabouts of Washington's Flammulated Owls still remain one of nature's many mysteries.

The Flammulated Owls paused their calls for a moment as the Barred Owl continued to call. But they soon began again. The drive to find a mate was stronger than their instinct for survival. Passion over survival. Love over fear. It was a lesson I would not forget.

Ukpik

Snowy Owls

There's an invisible ring around the globe at approximately 66°33′ N, known as the Arctic Circle. Here the sun does not set below the horizon on the summer solstice or rise above it on the winter solstice, leaving the earth in either perpetual day or night for months on end. Above this ring, trees look like shrubs, growing low to the ground over hundreds of years. In order to utilize their ability to make energy from sunlight, trees need warmth, a commodity that is lacking here. Just as snow-covered mountain peaks have a tree line, Earth has a bare spot near her poles where no trees will grow. Temperatures in this region have fallen to as low as -90°F and risen to as high as 86°F, yet the average daily temperature in summer doesn't exceed 50°F. Still, the Arctic, with all its extremes, provides a rich, specialized niche for those prepared to brave its bipolar swings.

The Snowy Owl is one creature that can survive the spectral ends of planet Earth. Special adaptations prepare this owl for life in the Arctic, hunting and surviving in months of darkness, snow, and ice. For all owls, hunting in near-total darkness is nothing special. Their long, tubular eyes are power-packed with black-and-white nerve cells, called "rods," and a tapetum lucidum—the same mirror-like structure that causes cat or deer eyes to shine in the dark. This special layer of tissue reflects light back to the rods a second time, to the point that the owls' nighttime world is brightened to what we would think of as cloudy daylight conditions. Combined with their specialized ears, this unusually good night vison makes all owls excellent nighttime hunters. But withstanding regular subzero temperatures is an exceptional adaptation, even for an owl. With this in mind, I traveled to Barrow, Alaska, to learn more about the extraordinary Snowy Owl.

Matt Larson, a research biologist from the Owl Research Institute (ORI) in Montana, walked two paces ahead of me with the swift stride of someone who hikes eight to ten miles daily. Every so often he stopped and scanned the open tundra with binoculars, giving me time to catch up. On this August day toward the end of the nesting season, we walked in the general direction of the nest where Matt had last seen the slow-to-develop female Snowy Owl chick we were checking on.

When he heard the sharp bark of a male Snowy Owl, he stopped, turned to me, and said, "This male is known to attack. Be careful." This was the chick's father, the one parent remaining to feed the chick until she learned to fly. Dumbfounded, I ducked, trying to make my head and shoulders less of a target while imagining the father's long black talons outstretched and coming for my head. An owl attack was not on my list of experiences I wanted to have today.

I'd met Matt a few months earlier near Charlo, Montana, when I'd gone out for the day with the ORI team to search for Long-eared Owl nests. Energetic, enthusiastic, and very knowledgeable about both owls and the surrounding habitat, Matt had taught me how to use my binoculars as a field microscope. He'd found an owl pellet on the ground near a roost tree, and when he picked it apart, he found the tiny jawbone of a vole. Then, by flipping his binoculars over and looking through the large end, he proceeded to show me the magnified jawbone. "See the second molar?" he asked, running his thumb along the minuscule jaw to point to the molar. "See how it's flat? There would be a slight curl if it was a *pennsylvanicus*," he said, using the scientific species name for the meadow vole. "This owl caught a *montanus*," he added, using the scientific name for the montane vole. "You can sometimes see their little tunnels running through the fields."

Through the makeshift microscope, I could see the vole molar clearly, and I was fascinated by this new trick. Now here in Barrow, I knew I would learn a lot from Matt about Snowy Owls.

"If he comes at you, don't just drop to the ground. He'll just adjust his attack," Matt warned as he gave me a brief How to Avoid Owl Attacks 101 lesson. "You have to wait until the last second. Kind of play chicken with the owl."

I stood, mouth agape and nodding, as if I understood through the rush of fear clouding my mind. I'd walked through meadows of resting grizzly bears and sat in an open Jeep just feet away from wild Bengal tigers, yet I'd never been so scared of an animal before. I couldn't believe I was afraid of an owl.

As we walked, I scanned the tundra for the fuzzy gray chick. She was well camouflaged against the background unlike her bright-white father, who sat barking a warning. At this point, late as we were in the season, the chick's mother was nowhere to be seen. She had flown away to replenish her own depleted reserves in time to survive the fast-approaching, long, harsh winter.

We walked over high-centered polygons—small tundra hills created by thousands of years of freezing and thawing of the water and ice within and below them. We hiked through gullies and puddles toward a point on the other side of a small lake. Every now and again, I was slowed by the boot-sucking mud at the bottom of these puddles, falling behind Matt, who walked with determination toward the chick that only trained eyes like his could spot.

Matt had learned from Denver Holt, the founder of ORI and one of the preeminent owl biologists in the country. Holt had been coming to Barrow each summer for the last twenty-three years to study the nesting behavior, diet, and habitat of Snowy Owls. This year he arrived to find what looked to be a very successful year for Snowy Owls raising chicks during what seemed like a population boom of brown lemmings, a favorite food of the owls. He found twenty-one active nests of Snowy Owls with clutches of seven, eight, and even nine eggs.

But somewhere between the eggs hatching and the owlets fledging, the population of lemmings crashed. By the end of Holt's visit, eight of the twenty-one nests had failed to fledge even one chick. Ten of the nests managed to raise at least one or two chicks, and only three nests were successful raising two or more. The owls were surviving on the nesting seabirds in the area, like Common Eiders, Surf Scoters, and the occasional Greater White-fronted Goose. Now Matt had arrived in Barrow to follow the progress of the chicks that had managed to make it so far this year.

Snowy Owls are very particular about where they nest, wandering around the Arctic in search of large populations of lemmings before they breed each year. Around Barrow the owls prefer brown lemmings; in other areas collared lemmings are the cuisine of choice. Either way, lemming populations are one of the most precarious populations of small mammals in the Arctic. These lemming species provide nourishment for many creatures around the Arctic Circle, like arctic foxes, wolves, wolverines, jaegers, and gulls.

Seemingly small differences divide collared lemmings and brown lemmings into separate species. Collared lemmings prefer dry ground to the brown lemming's favorite habitat—wet, marshy ground. In winter, collared lemmings turn white to camouflage against a backdrop of snow, whereas brown lemmings remain brown. The two front toes on a collared lemming enlarge in winter to form mini ice scrapers for tunneling through deep snow while the brown lemmings' toes do not. But brown lemmings have evolved a larger, thicker skull for bulldozing their way through ice and snow.

The diet of these two species is also different. Brown lemmings prefer to eat the succulent bits of cotton grass and sedge leaves that grow in marshy areas, while collared lemmings eat dry land plants, like willows and arctic dryad. Both lemming species go through regular cycles of population boom and bust every four years or so. Yet any manner of catastrophic change is likely to disrupt and interfere with the lemmings' delicate population balance, stretching the bust cycle to as much as eight years, leaving the Snowy Owls searching farther and wider for the lemmings they need to survive.

Snowy Owls arrive on their breeding grounds in late spring to assess the lemming population, find a territory, and mate. If no lemmings are found in one area, the owls move on, continuing to search the tundra for an area with an acceptable lemming population. Or Snowy Owls may decide not breed at all, even though they will eat a variety of other creatures, such as mice, red-backed voles, seabirds, and songbirds that breed in the Arctic.

The number of eggs Snowy females lay each year is also based on the lemming population. In a boom year for lemmings, each Snowy Owl female may lay eight to eleven eggs. In an average year, she may produce only two to four eggs. The owls make a calculated guess in the spring based on the lemming population at the moment, which may not hold through the summer. If the snow melts too quickly or if it rains too much in the spring, lemmings move to higher ground, making it look to Snowy Owls as if there are more lemmings than there really are. That's one reason why nests fail altogether in some years. The chicks that had survived this summer in Barrow seemed to have done so only by the sheer determination and skill of their parents.

On this particular day, as Matt and I searched for the slow-developing Snowy Owl chick from one of the successful nests that Matt had been monitoring, the father owl was not happy about it.

"I'm going to catch this chick," Matt said. "Come along at your own pace."

I walked along as slow as molasses on a winter day as Matt closed in on the chick and the male, who looked like he meant to fly right at us if we didn't stop. I tried to decide whether or not to risk attack. I picked up my pace, moving as fast as my wader-covered legs would go, feeling simultaneously worried for Matt and scared for myself. Should I go forward and help or linger behind and protect my own head? I figured Matt knew what he was doing. Still, I was apprehensive. Most of the owls I'd studied to this point were small and not to be feared. A large angry father Snowy Owl intent on protecting his remaining chick was something I hadn't bargained for.

The Snowy Owl (*Bubo scandiacus*) is the heaviest of North American owls, weighing in at four to six pounds, as much as two pounds heavier than the owl closest in size, its cousin the Great Horned Owl (*Bubo virginianus*). Most of this weight comes from thick fur-like feathers that cover everything from the Snowy's feet to its face to the tip of its black beak. This heavy down jacket allows it to maintain its internal body temperature of 95°F to 104°F, even with an air temperature of -58°F. Below that point, they rely on other biological factors, such as stored fat reserves and behavioral mechanisms, like roosting out of the wind and more frequent hunting and feeding, to provide the energy necessary to help keep them warm, even when it's -80°F.

As an adaptation to ever-changing prey availability, Snowy Owls are nomadic. They circle the North Pole in the northern latitudes, stitching together artificial human boundaries. From 1999 to 2001, a study done by biologists Mark Fuller of USGS, Denver Holt of ORI, and Linda Schueck of Boise State University utilized the then-new technology of the Argos satellite telemetry system to track four female Snowy Owls. The four females were captured and radio-tagged in Barrow during the nesting season. Each was fitted with a small backpack, secured with a Teflon ribbon and containing a radio transmitter, and identified by the number on her leg band. When the owls were released, every four days a signal was sent from the tag to polar-orbiting satellites, which in turn forwarded location estimates to the researchers, allowing the team to use their desktop computers to track the wide-ranging movements of the Snowy Owls to within about a half mile.

One of the owls, #57, was tagged in July 1999 and tracked for two years. After successfully raising her chicks, she left the nesting grounds on August 20, 1999. From Barrow, the northernmost US city, on the coast of Alaska, she first headed inland about a hundred miles before turning north and heading out over the Chukchi Sea ice. There she hunted sea ducks in polynyas—patches of open water that form where warm-water sea currents upwell from below. She settled on the Alaska coast about one hundred miles west of Barrow on October 19 for her first extended period, about seven weeks. On December 8, she flew west over the Chukchi Sea toward the Chukotskiy and Kamchatka peninsulas of Russia, where she wandered for four months during the darkest and coldest time of the year, flying as far south as 59°N, yet never staying longer than two weeks in any one place.

From March 27 to April 18, 2000, #57 settled again, this time near Chuan Bay, along the Russian coast of the East Siberian Sea, about 250 miles west of Wrangel Island, a known island hot spot for nesting Snowy Owls. She then moved another five hundred miles or so farther west, settling less than one hundred miles from the coast from April 27 to August 24, 2000, where she most likely nested, nearly twelve hundred miles from where she had nested the year before after a nine-month, thirty-two-hundred-mile journey.

In late August, #57 left her Russian nesting grounds, this time flying east. Her first extended stop, from November 27 to December 13, was back in Chuan Bay. Starting on December 20, she spent the next four months in the middle of the Chukotskiy Peninsula. On April 2, 2001, she left her wintering spot and headed east toward Barrow, where she'd nested in 1999. However, about two weeks later she passed by Barrow.

It's possible that she was searching for a suitable nesting site but, not finding one, did not stay; in Barrow during the summer of 2001, the lemming population was very low. No Snowy Owls nested in Barrow that year. Instead, #57 continued east until she settled on Banks Island in Canada, a well-known nesting area for Snowy Owls, from May 20 until July 22. That means that this female Snowy Owl made a journey of over sixteen hundred miles in six weeks, ending up approximately one thousand miles from her 1999 nesting site in Barrow.

Over the two years #57 was tracked, she traveled nearly five thousand miles, most likely nested three consecutive years (including the year she was tagged) thousands of miles apart, and crossed three international borders. The other three owls tagged in the study showed similar wide-ranging movements. These movements are not considered migratory in nature, however, because the owls don't return to the same place each year, nor do all the owls go to the same place. So while Fuller, Holt, and Schueck's study does shed some light on where Snowy Owls go, and while it seems that these wide wanderings must almost certainly be at least partly food-related, scientists still do not know exactly why Snowy Owls go where they go.

In March 2014, the International Snowy Owl Working Group (ISOWG), with members from the United States, Canada, Russia, Norway, Sweden, France, and Germany, announced that the Snowy Owl population—at one time thought to be 290,000 to 300,000 individuals worldwide, giving the species a status of "least concern" as classified by the International Union for Conservation of Nature (IUCN)—was considerably overestimated. ISOWG said a better estimate puts the population at around 14,000 breeding pairs, or just 28,000 individual Snowy Owls, in the world. With increasing challenges for Snowy Owls posed by global climate change, ISOWG recommended that Snowy Owls be included on the IUCN Red List of Threatened Species and on the Arctic Migratory Birds Initiative's list of priority species. While the IUCN recognizes that the Snowy Owl population is decreasing, the status of the species remains unchanged for now. With all that we still don't know about Snowy Owls combined with the unknowns of climate change, it will certainly take international effort and cooperation across boundaries to keep this majestic species from slipping further toward endangered status.

Snowy Owls seem to get the solo travel bug early, leaving the nest for far-off destinations on their own. From one nest on Victoria Island between the Arctic coast of Canada's Northwest Territories and Nunavut, three Snowy Owl chicks each chose a different path, each following its own inner nomad. One flew west to the east coast of Russia, another went south to Hudson Bay, and one headed to southeastern Ontario, bordering New York and New England.

In their first winter, chicks often head south, where hunting may be easier for first-year owls. In boom years, also known as "irruption years," Snowy Owls begin to appear as if by magic in places they aren't usually seen. In the irruption year of 2011–2012, Snowies were seen in Seattle—on rooftops in urban neighborhoods like Capitol Hill and Ballard, and in the wide-open spaces of Discovery Park. There were also thirty Snowy Owls that year around Boundary Bay on the Washington–British Columbia border. And there were more than ten along the Columbia River on the Washington-Oregon border. It was so unusual that in February 2012, *NBC Nightly News* sent a reporter to Ocean Shores, Washington, to cover the story.

The Pacific Northwest wasn't the only region blessed with sightings of these rare arctic visitors that year. "There are so many across the country, everywhere, by the thousands," Denver Holt told a reporter about the Snowy Owl irruption for the story "Bird-Watchers Revel in Unusual Spike in Snowy Owl Sightings," published in the *New York Times* on January 22, 2012. "It's unbelievable," Holt continued. "They are being seen from Boston, to the Great Lakes, the Ohio River Valley, Kansas, Vancouver and Seattle." On eBird, an online birding website jointly run by the Cornell Lab of Ornithology and the National Audubon Society, birders reported real-time Snowy sightings as far south as Arkansas and Texas. Even Honolulu, Hawaii, had one Snowy Owl, the first ever recorded in that state. That Snowy Owl made a nearly three-thousand-mile flight, landing at the Honolulu International Airport on Thanksgiving Day 2011. However, fearing collisions with airplanes, officials from the USDA's Wildlife Services Division promptly shot the owl. "It's the first ever in Hawaii and they shot it!" said Holt to the *New York Times*, astonished.

Snowy Owls often choose airports as landing sites in unfamiliar territory, preferring wide-open flat spaces reminiscent of arctic tundra rather than the deciduous and fir forests of the Lower 48. Norman Smith, director of the Mass Audubon's Blue Hills Trailside Museum, has been trapping and relocating Snowy Owls from Logan International Airport in Boston since 1981. He regularly removes six Snowies a year, but in another irruption year in 2013–2014, a result of a booming collared lemming population and successful nesting season in northern Quebec, Norman Smith relocated 120 Snowy Owls from the airport, obliterating the previous record of 43 set in 1986.

The 2013–2014 irruption also afforded ornithologists another opportunity to study this majestic owl. Since 1999, when Snowy Owl #57, tagged in Barrow, Alaska, journeyed to Russia and then back to Canada with her radio tag relaying rough location estimates every four days for two years, technology has blossomed. During the 2013–2014 Snowy Owl irruption, a research project dubbed Project SNOWstorm was conceived and publicly funded through a crowdsourcing campaign. Researchers were able to capture twenty-two Snowy Owls in the Lower 48, tag them, and track their nearly real-time movements with new solar-powered GPS units that relayed precise latitude, longitude, and altitude coordinates as fast as every thirty seconds through cell-phone-tower transmissions.

Once the owls moved back north into the Arctic, cell phone towers became scarce. But with the solar-powered transmitters and improvements in data storage capacity, the transmitters continued to record and store data. In 2015, five of the tagged owls traveled south again for the winter, transmitting data from the first cell phone towers they encountered and giving scientists their first look at the precise summertime movements of Snowy Owls.

Through the Project SNOWstorm website, the general public can also follow the daily lives of these charismatic owls. As the Arctic warms through global climate change, we will be witnesses to both the challenges and successes of these most beloved owls.

Everywhere they go, Snowy Owls draw throngs of bird-watching paparazzi and other excited onlookers. During the irruption of 2013–2014, I was among the enthusiastic public, hoping to see my first Snowy Owl. On a tip from Norman Smith, I traveled to Salisbury Beach State Reservation, a state park in Massachusetts. "There've been a couple of juvenile males sitting at the entrance to Salisbury State Park all winter," Smith had told me. "That's a good bet."

I took that bet and arrived at the park on a rare sunny afternoon to scores of cars parked alongside the road and people standing between them, staring through binoculars. From years of watching wildlife, I knew the signs. It was the kind of spectacle you would expect to see when a bear walks alongside the road in Yellowstone National Park. I hoped they were all trying to get a glimpse of the same rare arctic visitors I was here to see.

I lifted my binoculars to my eyes and scanned the snow-covered marsh but saw nothing at first glance. I studied the people to my right in an attempt to determine where they were looking. They were already walking away, but I noticed a man with a camera and huge lens mounted on a tripod walking out onto the icy marsh, so I watched to see what he would do. He pointed his camera at a brown tuft of sea grass. Was he looking at a Snowy Owl? I couldn't tell, and I was beginning to wonder if another interesting species was the object of such curiosity.

The people standing next to me seemed to be actively watching something, so I walked closer and asked, "Are you looking at a Snowy Owl?"

"Yes," said the woman. "It flew behind that little tuft of grass at the edge of that hill. It's a bit hard to see."

Sure enough, I could just make out the Snowy's head between the stems of dry beach grass. I watched for a while, hoping the owl would move so that I could get a better look and a good photo. He didn't, so I walked around the corner toward the man with the camera on the frozen marsh. I thought about venturing out to his spot, but I didn't want to disturb the owl.

Winter is a harsh time of year for almost any animal to survive and this year, with a polar vortex, was harsher than usual. I knew that any energy the owl expended flying away from overzealous humans was that much less he would have to stay warm and find food. It's always best to give animals their space, which is quite a bit larger than our own personal space. So I stayed on the road, watching through binoculars as the regal owl held court on the frozen marsh that winter day.

As I stood watching, people stopped to talk. "Sometimes he sits up there," said one man as he pointed to a nearby electrical pole. He was a local wildlife photographer who had been regularly visiting the park to photograph the two Snowy Owls there. He'd spent hours over many days studying the owls' behavior to get the best photos possible. "There's usually another one on that side," he added as he searched the field on the opposite side of the road. "I don't see him now. Wonder where he's gone."

"I've seen them sitting on the pink house at the edge of Parker River on Plum Island," another birder told me, then gave me directions to find the place.

Some of the other people passing by were also seasoned birders; others were simply curious about the Snowy Owls they'd been hearing so much about. Many paused to tell the story of their sightings, as if sharing the experience with a fellow enthusiast would somehow make their extraordinary encounter seem more real. I talked with these people like I had known them for years. I didn't know their names or anything about them other than they shared my delight in spotting Snowy Owls. Nothing else seemed important at what seemed like an impromptu festival centered on these spectacular snow-white birds. I gathered their stories and information, fitting together the pieces of my owl puzzle.

Then, for those just arriving, I became the expert. "Do you see the owl?" they asked. "What are you looking at?" And I told them what I knew. When one mother with her two young daughters walked by looking for the "Harry Potter owl," I handed them my binoculars so they could get a closer look.

The next day the weather changed to a cold, persistent drizzle. No other onlookers joined me, and I sat comfortably in my rented car, scanning the marsh for any sign of the owl I'd seen the day before. I finally found the Snowy sitting atop his favorite electrical pole, intently watching the marsh for any moving thing. I also found the other owl in his territory across the road, just like the photographer from the day before had told me. At one point, I found myself sitting between the two Snowy Owls, one on each side of the road, and I felt like an owl as my head swiveled, unsure of which one to watch. Unlike the owl, however, sometimes I had to turn my whole body in my seat. The owls turned nothing but their heads for long periods, as they sat scanning the marsh for potential prey.

Because owl eyes can't move as our eyes do, owls rely on turning their heads to see. The fourteen vertebrae in an owl's neck, compared to our seven, allow their heads to turn 270 degrees without cutting off the blood's circulation to the brain. The holes in an owl's vertebrae through which arteries pass on the way to the brain, called the "transverse foramina," are ten times larger than the arteries, thus giving the critical blood vessels plenty of room to move as the owl's head turns and its body remains motionless.

I was fascinated by every move of the owls that day, as I was allowed the privilege of witnessing the natural behavior of these two Snowy Owls, undisturbed. Unlike the brief interactions with Saw-whet and Flammulated Owls I'd had up to this point, I observed the Snowy Owls for hours in daylight. As I watched them, I gained a greater sense of what it means to be an owl: their movements, catlike in their patience and watchfulness; their fearless, aloof apex-predator demeanor; their almost magnetic draw. Yet even as I gained a greater understanding of Snowy Owl behavior, much mystery remained. I wanted to learn more. To do this, I wanted to see Snowy Owls in their natural nesting habitat on the arctic tundra.

In August, six months after my first Snowy Owl experience in New England, I left sunny 88°F Seattle and hopped on an Alaska Airlines flight to Barrow. To say it was a shock when I stepped off the plane to 27°F with windchill would be an understatement. But I'd come prepared with many layers.

The afternoon I arrived, Matt Larson picked me up at my hotel on a four-wheel ATV with a folded caribou hide across the back to make a seat. While I'd been expecting a truck, I was up for the adventure and climbed aboard for a frigid ride. I learned quickly how ill prepared I was as the wind from the speeding ATV joined the stiff breeze off the Chukchi Sea, blowing straight through my layers and taking my breath away. Luckily, the cozy caribou hide helped to keep at least some part of me warm on this summer day in the Arctic.

We drove out of town heading west on the network of dirt roads, passing teenagers playing basketball in shorts and T-shirts outside weathered houses with windows blocked by tinfoil to keep out the midnight sun.

Other curious details caught my eye as we made our way out of town—all representations of the subsistence lifestyle in Barrow—a severed moose head, antlers and all, lying on the roof of one house; sealskins stretched to dry outside another; strips of ten-foot-long fringed baleen leaning against yet another house; and a whole bowhead whale skull decorating the outside of one business.

We passed a sign pointing in many directions with a painted Snowy Owl cutout on top. It marked the distance to the North Pole (1,250 miles), South Pole (11,388 miles), New York (3,380 miles), London (4,114 miles), and Seattle (1,960 miles). Now I really was the farthest out "in the middle of nowhere" that I had ever been. There are only two ways into or out of Barrow, either by plane or by boat, and that is for only part of the year. Food we take for granted, like milk, is flown in from Anchorage or barged in through the Bering Strait; in Barrow, milk costs as much as twenty dollars a gallon.

On the outskirts of town we passed another sign, a large blue one with an even larger Snowy Owl carved in relief on one end, which read "Ukpiagvik, the place where we hunt Snowy Owls." This ancient site, once a village of igloo-shaped sod homes, has been here for over two thousand years, highlighting the long tradition of humans using this unique ecological niche, surviving and thriving off the arctic land, sea, and ice. Here Snowy Owl eggs were gathered, owls eaten, and their feathers used on arrows. Snowy Owl talons were worn at the waist on a belt, which was believed to encourage the development of strong fists. Nothing was wasted.

Ukpik or *Ookpik*, as the Snowy Owl is called in the Iñupiaq language, is featured in several morality tales, which reveal the Snowy Owl's vanity and pride. In one story, Ukpik fell in love with a beautiful ptarmigan woman, a little white bird called Aqilgieq. Ukpik wished to marry the ptarmigan, and in a jealous rage, he killed her husband, whom she loved very much. Ukpik sang of his love to the little ptarmigan woman, but she could only cry for her lost husband. Aqilgieq refused the owl, calling him ugly and asking, "Who would want to marry you?" Ukpik, thinking himself very beautiful, flew off in anger.

In another story, Ukpik was very hungry, so he stalked and captured Lemming. Ukpik, so proud of his skill and dominance as a hunter, began to dance. Lemming encouraged his dancing and complimented him on his skill and grace. Ukpik became so enthralled with his own dancing that he forgot to pay attention to Lemming, who slipped away unnoticed. The Snowy Owl is also involved in the Inuit story of why ravens are black. Raven and his friend Owl were playing a game one night but quickly grew bored, so they decided to paint one another. Raven did a beautiful job, painting the tips of Owl's feathers black. Owl was so happy with Raven's painting that he gave his friend a new pair of fur boots called "mukluks." Raven was so happy with his new boots that he danced and danced, and he wouldn't stand still long enough for Owl to paint him. Owl finally got so angry that he dumped the bucket of black paint on Raven, who is black to this day.

Looking at the large Snowy Owl sign in Barrow, I realized that the Inuit people's long association with the Snowy Owl pointed to the fact that Ukpiagvik has been a Snowy Owl hot spot for more than two thousand years.

Another mile or so out of town, Matt parked the ATV within sight of two Snowy Owls sitting out of the wind, hunkered down behind two high-centered polygons. Behind one of the owls in the distance, the Wiley Post–Will Rogers Memorial Airport lay low and shimmering against the horizon. Behind the other owl, the open tundra stretched south to the Brooks Range, the mountain chain that defines Alaska's North Slope. About three hundred yards of tundra lay between me and the owls. The yellow-gray tundra grasses grew beside tufts of white cotton grass that looked like miniature Truffula Trees, the trees created by Dr. Seuss for *The Lorax*.

The Snowies sat undisturbed and unruffled by our presence, just as I had seen them do in New England. Yet the tiniest movement attracted their attention. Their heads swiveled as if on pins, turning from side to side toward anything that could mean their next meal. And here I stood in the Arctic, in a wilderness unlike any other I'd ever encountered, watching the object of my quest—an owl like no other.

These white arctic wanderers living on the edge of all that we commonly know seemed more closely related to my own wandering soul than any other owl. Yet their lives and what they endure, in what seemed like perpetual winter to me, were far from what I could fully comprehend. With the naked eye, I could see their white feathers lightly touched with the black tip of Raven's brush. Through my binoculars, I could see the bright-yellow eyes that had seen so many faraway places I've wanted to see. But I was curious about the maps they carry within that steer them toward the next lemming hot spot. I could imagine the long black talons grasping quick prey. But I was still puzzled about how Snowy Owls so often successfully predicted that the lemming population would last through the summer, long enough to raise their chicks. Yet here they lived, in the extremes of planet Earth, wild and free in the Alaska wilderness, where even the flowers are furry.

The two owls we watched were male Snowies that had not found mates for the year. There were no nests in this area, but there were always a few owls here, even when the owls weren't nesting. I could tell the male from the female Snowy Owl by the whiteness of the feathers. Adult females and juveniles are more heavily barred on the wings, chests, and head. As male Snowies age, their feathers turn whiter. It would seem that they turn white for camouflage, but in the summer against a tan tundra, they stand out like a beacon in the night. Some scientists think that Snowy

Owls may be able to see ultraviolet light and that their white feathers are a signal only other Snowy Owls can interpret.

Females are also noticeably larger than males, and this reversed sexual size dimorphism is very clear when Snowies sit side by side. During the many months of brooding, the female is entirely dependent on the male to bring food. She incubates the eggs, and then the hatchlings, until they are big enough to thermoregulate, generating their own body heat. Only then may she join in the hunting of lemmings or small birds needed to feed the growing chicks. She will leave the nest for good once the chicks have gained the ability to fly. The Snowy male stays around for a few more months to feed the chicks as they grow and learn to hunt on their own.

By the time I arrived in Barrow, the female Snowy Owls had left their nests and most of the chicks had fledged, except for one slowly developing female with the angry father—the chick Matt and I would look for. Matt picked me up again the next day on the ATV. This time he brought hip waders for me to wear on our hike. As we sped along Cakeater Road outside of Barrow, Snowy Owls dotted the gray-brown tundra that surrounded us. I could see them gleaming like beacons, even through the protective goggles I wore. Matt pointed to each owl near a nest, and I envied the owls for their down jackets, which allowed them to sit comfortably in the cold arctic wind.

Matt parked the ATV on the side of the dirt road about ten miles out of town, and we gathered our gear for the hike. I carried binoculars and a camera. Matt carried a small backpack with owl banding equipment and a notebook filled with pages of data collected on each owl. He would document the fledging of Snowy Owl chicks from twenty active nests around Barrow that year for ORI's Snowy Owl project.

Just after the days of twenty-four-hour sunlight had ended, Barrow was anything but summery. Yet I found myself squishing across melted and surprisingly springy tundra. Dressed in five layers plus hip waders and a thin wool hat, I was left feeling like a knight in camouflaged armor. If not for the constant wind that took the temperature from a somewhat reasonable 37°F to a breathtaking 21°F with windchill, I would probably have been covered by mosquitoes, a fact that I did not fully appreciate in the moment. I was more worried about long black talons piercing my thin hat and my skull beneath it than I was about a few mosquito bites.

Matt swiftly moved toward the slow-to-develop chick. When he reached the chick, it began to hop away, its undeveloped wings outstretched. I stopped and held my breath as the father owl took flight. Then I took a few steps, moving faster than before, one eye on Matt's progress, the other on the flying male. Matt swung around behind the chick, chasing her away from the marshy area beside her, trying to keep her from jumping into the freezing water in her attempt to escape. Wet feathers could be life threatening. With one swift movement he leaned over, put one hand on her neck behind her head, and dropped to his knees to contain her wings.

Matt tried to open his pack one handed, so I knelt down to help, dropping my camera and binoculars beside the polygon, where we crouched out of the wind, knowing time was of the essence. "What can I get you?" I asked.

"Here. You hold her," he said, passing the chick to me. "Keep her wings against your chest and hold her feet like this." He showed me how to turn my hands with my thumbs in front so that I could support her weight and hold her feet tight in my hands. All of a sudden I was cradling a baby Snowy Owl, the biggest owl I'd ever held. I was nervous. Was I holding too tight? Would her fragile baby owl bones snap in my grip? I tried to calm my beating heart. My legs bent awkwardly underneath me and began to cramp. I tried to move without causing unnecessary torture for the frightened chick.

As Matt searched his pack for his notebook full of Snowy data, he seemed unhurried. I relaxed from his example. I settled down to hold this baby Snowy as safe and secure as if she were my own child. I tried to take in all the details of this extraordinary encounter with this wild arctic owl beside this small hill on the tundra.

Fluffy gray down mingled with smooth black-and-white adult feathers. Around the chick's bright-yellow eyes, tiny white feathers began to form the facial disk that would one day funnel tiny sounds to her ears. Gray baby down feathers still covered her large feet with three-quarter-inch black talons.

I glanced around, looking for the male on the wing, as Matt read the number on the chick's leg band. The male Snowy Owl was nowhere to be seen. Matt found the chick's page in his notebook, carefully measured the length of her tail and her wings, and recorded the numbers. "She's growing," he said, noting the difference from previous measurements. But why was she slow to develop? It was most likely not due to her birth order. *Was it a result of the lack of food that the Snowy Owls had encountered midseason?* I wondered.

Quickly tiring of this unwanted poking and prodding, the chick began to squirm in my arms. She managed to find some bare skin on my right hand, leaned over, and pinched the soft flesh where the thumb connects to the hand. "Ouch, you bit me!" I cried, knowing full well I deserved it for restraining this wild creature. Then her left wing unfurled, exposing the chick to the cold arctic wind.

"Ah, you're feisty," Matt said to the chick, as he tucked her flailing wing back in. "I like that." Then he borrowed my camera to take photos of her face and wings for later study of how well she was developing. "See how her feathers are still encased," he said, stretching out her left wing. The chick's primary feathers were still covered in the drinking-straw-like sheath in which they grow. Normally, at about sixty days, this chick should have full flight feathers and the ability to fly, but she was behind. And without the full benefit of her adult feathers, flight was impossible for this chick.

After collecting all the measurements necessary, Matt packed his gear, and I handed the baby owl back to him. I noticed the father owl sitting on the tundra several hundred yards away, watching our every move, having decided not to attack us on this day. Matt then placed the baby on the ground, facing away from a boggy patch near the polygon, again not wanting her to hop into the water as soon as he released her. He held her with his hand behind her neck, just as he had caught her. When he let go, we walked away quickly so that she wouldn't feel she had to hop to safety. We stopped to look back once we were several hundred yards away, only to see the baby owl splashing in the water we had hoped she would avoid.

"Will she be okay?" I asked Matt.

"Yeah," he said. "She can hunker down behind a hill out of the wind. She'll dry pretty quick."

The baby hop-swam farther away until all I could see were two yellow-and-white daisy eyes peeking at us through the tundra grass. *Good luck, little owl*, I silently wished. *Have a good life.*

A thick fog had rolled in off the Beaufort Sea while we'd been working with the chick. We'd walked only a mile or so from the road. Matt regularly walked eight or ten miles, alone, for his daily nest checks. I knew he carried a GPS in his pack, but I realized for the first time how easy it would be to get lost in this wilderness without trees.

"Do you know where we parked?" I asked Matt.

"I have a good idea," he said. "We'll find it, eventually."

We walked on across the tundra, with Matt naming the plants I pointed to, like five-petaled saxifrage flowers, pointed ivy-like leaves of coltsfoot, and antler-shaped reindeer lichen. We also saw some reindeer droppings. Denver Holt insists on his assistants being well-rounded naturalists, a challenge that Matt had taken on with gusto. Knowing as much as possible about the Snowy Owl's environment, the plants and animals that share the owl's habitat, can give clues that help tell the story about the Snowy's success or failure from year to year.

"Want to see the nest?" Matt asked as we neared the nest where the little chick had been raised.

"Yes!"

"It's right over here." Matt pointed to a high-centered polygon. The small hill stood only about knee-high but gave a wide view of the surrounding tundra.

"The female scratches out this shallow bowl with her talons, then lays her eggs," he explained. I knew that the mother would then begin to brood each egg as soon as it was laid, unlike other birds that begin to incubate the eggs only once a full clutch is laid. Once the eggs hatched, there could be as many as three weeks between the oldest and youngest chick with the biggest and strongest chicks having the best chance for survival.

"That's the nest?" I said, staring at the indentation in the hill before me. It was barely bowl shaped, with hardly any lining at all. I would have expected it to be full of warm downy feathers, but instead there was a bit of dried grass, the gray feather remains of a songbird meal, a few scattered baby feathers, a few mouse-size gray owl pellets, and the chalky white bird poop that owls produce. This was the nest the mother Snowy Owl had created to raise her chicks before she left them forever, having given all she could.

A deep yawn overcame me then as I pulled my boot out of the mud yet again. "It can be tough physically and emotionally being up here alone, walking out on the tundra," Matt had said earlier about being the only biologist from ORI still here checking nests. Now I understood what he'd meant. Staying warm in the biting wind, fighting to move against the tough plastic of my hip waders, and struggling to pull my feet out of the mud, combined with my nervousness about avoiding the angry father and my excitement at holding the snowy chick, had all taken a toll. I was exhausted to my bones. I could also appreciate, in this moment beside the empty nest, the emotional distance biologists commonly try to keep from the animals they study.

"Wow! I need a nap," I announced. "How do you do this every day?"

"I take lots of tundra naps. They're the best. I get down out of the wind behind a polygon, like a Snowy Owl. I lie down in the sun, close my eyes, and hope I don't wake up with a polar bear licking my face. Fifteen or twenty minutes after lunch and I'm good."

It sounded appealing, but on this day, even with all the polar bears in the area about a mile out of town munching on a whale that had washed ashore, the sun wasn't shining and I preferred to nap in a warm cozy bed.

By the time we reached the ATV, the fog had cleared as quickly as it had settled. On the way back to town, we stopped once more to check on three chicks Matt spotted from the road, their daisy eyes against the brown tundra giving them away. The chicks flew as Matt walked toward them, and he didn't chase them. We watched through binoculars as one of them tore bits of flesh from the remains of a sea duck it held.

We climbed aboard the ATV, and I was once again on the warm caribou-hide seat in back. I searched for Snowy Owls dotting the tundra as we sped toward Barrow—the place where I'd now hunted for Snowy Owls.

Two weeks later, Matt looked in on the little chick again. "I checked the day I left and found the chick to be nearly fledged. I didn't catch her, although I probably could have, but she was flying short distances and looked to be very close to fledging. Just three weeks behind schedule," he told me. This was good news. I was happy to hear the baby owl was on her way toward surviving her first year.

Back in my warm home near Seattle, I felt grateful for the once-in-a-lifetime experience I'd had with wild Snowy Owls, circumnavigators of planet Earth.

One of my favorite qualities of wild species is just that—they are wild. They are free to move about the earth in their own time for their own purpose, a luxury that we as humans rarely experience.

They come and go as they please, showing up sometimes where we least expect them.

So it was when three months after my owl adventure in Barrow, I found myself standing on the Edmonds Pier, about a half hour drive north of Seattle and only ten miles from my home, watching a Snowy Owl sitting calmly on the harbor breakwater.

Dark barred feathers on the owl's head, breast, and wings showed that this was a young bird, flown south probably for its first winter. I thought of the gray Snowy chick I'd held in the Arctic and realized she would be grown by now. I imagined strong, fully developed wings carrying her south for the winter.

The white daisies of this juvenile Snowy Owl's facial disk now blended into the white feathers of an adult, surrounding bright-yellow eyes that carefully watched the Western Grebes and Pigeon Guillemots floating on Puget Sound. Two local crows pestered the owl like they pester Bald Eagles. The owl turned its head, body still. I checked its legs for any bands it might have. I did not see any.

It was cold and rainy, yet not as cold as a summer day in the Arctic. And I was not alone, as I had been on the rainy day I saw the two Snowy Owls in New England. I was surrounded by birders and wildlife photographers, all trying to catch a glimpse of this surprise arctic visitor near Seattle. I smiled at this wild life.

Rarely Spotted

Northern Spotted Owls

Perhaps the most iconic of all owls in the Pacific Northwest is the Northern Spotted Owl (*Strix occidentalis*). I was excited to look for these owls in the wild, in the ancient trees where they live, before they disappear altogether. Sighting one seemed to me to be as epic, and just as improbable, as sighting a unicorn. Yet from all I'd heard from the biologists called "hooters"—for the hooting call that they make either electronically or with their own voices to determine if Spotted Owls are nesting in the area—these owls were still out there. Still, even the experts were finding only one or maybe two nests in their survey areas each year. Tracking them would be more of an adventure than I could yet imagine.

Through the grapevine of owl biologists, I contacted Stan Sovern, a Forest Service biologist who has been monitoring the Spotted Owl population in Washington State, both on the Olympic Peninsula and in Central Washington, for the past twenty-eight years. He invited me out for an afternoon check of one active nest site. I met Stan and Margy Taylor, another longtime Forest Service hooter, at the Cle Elum Ranger Station. Then we drove east along I-90 to the forest edge, where the habitat abruptly turns to shrub-steppe, as if a line had been drawn through the state.

Just before we took the Taneum Creek exit off the interstate, I could see fields full of tall windmills turning in the breeze blowing down from the eastern Cascade slopes. *There are spotted owls here?* I wondered.

We then turned and headed west into some of the last remaining old-growth forest in Central Washington. It was here that we met William Meyer, a biologist from the Washington Department of Fish and Wildlife, who knew Margy from the days they had counted Spotted Owls together on the Olympic Peninsula. He now specializes in restoring beavers—once an integral part of the landscape—to their natural habitat and reclaiming watersheds that have dried up since beavers were removed from the land.

We wound around on back roads and dirt Forest Service roads. I was properly sworn to secrecy about the exact location of the nest and instructed not to return on my own. *No problem there*, I thought. I couldn't find my way back if I tried. We hiked along a recreational trail that had been heavily used by motorbikes and ATVs. I wondered how the elusive Spotted Owl lived among all this activity. Little did I know we were still far from the nest site. We crossed a little bridge and then headed off-trail, straight up a steep slope. It was slow going. Hiking uphill would not have been a problem for me if it hadn't been for climbing over all the downed dead trees through tangles of short shrubs.

Spotted Owls are fairly particular when it comes to choosing territory. While other owls easily use a variety of open fields or forest edges, Spotted Owls prefer dense old-growth forests with several very specific characteristics present. The more of these features a forest can provide, the better. A closed canopy for protection from avian predators like Northern Goshawks and Great Horned Owls is highly valued. Standing dead trees, called "snags," provide homes for their favorite prey, flying squirrels, and dead decaying logs provide great habitat for the squirrels' food. Different aged trees and shrubs, called a "multilayered canopy," and dwarf mistletoe infestations provide suitable Spotted Owl nesting habitat.

Once Spotted Owls choose a territory, unless some major event occurs, such as a fire or clear-cutting or the death of a mate, they tend to stay on the same land for the rest of their lives. I once heard a biologist telling the story of a pair of Spotted Owls sitting in the last remaining tree in the middle of a clear-cut, as if wondering what had happened to their neighborhood.

In the winter months, they do not migrate as some owls are known to do. Instead, they expand their home range as food resources are depleted, while their core habitat remains the same. Because they live so far from the beaten path, humans rarely see Spotted Owls, as I was learning.

"From a Spotted Owl's point of view, tangled and complex are very good things," Margy explained as we continued to climb. I huffed and puffed as I hauled myself over large fallen trees and ducked under low branches, hiking a few steps at a time up the steep slope. I gained a new respect for these biologists, who hike these forests daily, giving their careers to studying and counting these remote, mysterious owls.

"Tangled and complex" describes not only the Spotted Owls' habitat, but also other aspects of their life, including their diet. Spotted Owls rely most heavily on their favorite prey, the nocturnal northern flying squirrel. Flying squirrels are rather small as far as squirrels go, measuring only about ten inches from the tip of their long whiskered noses to the tip of their flat tails—just the right size for owl prey. With the use of their flat tails as rudders and the skin flaps that hang along their sides between their front and hind legs, these cinnamon-colored squirrels can soar between trees for up to 150 feet or more, looking much like wingsuit flyers leaping from mountain ledges.

Flying squirrels live solely in old-growth forests, eating false truffles, a type of fungus that grows only in undisturbed decaying woody debris on the old-growth forest floor. These truffles spread through the soil attached to the roots of trees, gathering water and nutrients and fixing nitrogen into a usable form that the fungus can share with the trees. The trees in turn provide sugars and carbohydrates that the fungus needs to thrive. This symbiotic relationship is furthered when the fungus fruits into the truffles that the flying squirrels depend on as their favorite source of nutrition. The squirrels spread the spores of the truffles around the forest, far and wide, as the spores pass through the squirrel's digestive system intact. When a squirrel is caught and eaten by a Spotted Owl, the spores can spread further through the dense forest in the pellets the owls expel. Thus, for the health of the forest and the good of all, each species plays its part as prey or predator in this cycle of well-orchestrated forest theater.

As I reached the top of the slope, my heart pounding and breathing heavy, Margy stood looking through binoculars toward a large cedar with a clump of dead branches on its left side—commonly called a "broom," which had likely formed from a dwarf mistletoe infestation. Another dead tree leaned against the cedar just above the broom. Tangled and complex. An owl flew toward us, landing on a low-hanging branch very near to where Margy stood. Now my heart pounded for a different reason as I stood no more than twenty yards away, in the presence of the mythical Spotted Owl.

Two round black eyes stared at our small group. White semicircles between the owl's eyes met in the middle just above his yellow beak, making a white *X*. A thin dark-brown line of stiff feathers outlined the owl's lighter-brown facial disk. White spots, for which the Spotted Owl is named, adorned his brown head, chest, and wings. He seemed gentle and unafraid of us. "They used to follow us down the trail as we hiked," William said, remembering his days hooting in the Olympic Mountains.

I could not believe our luck. Here was a Spotted Owl sitting so close, watching us. However, Margy soon explained that this wasn't the result of luck but years of hard work. Both she and Stan knew this owl. Stan had observed this male Spotted Owl since he was born in 1998. He had tagged him with his uniquely colored leg band for easy identification when he was a chick in a nest two miles from here. The male first mated as a two-year-old and successfully raised two chicks each year in 2000, 2001, and 2003. The pair moved to this nest site in a nearly two-hundred-year-old tree in 2004. After his first mate died in 2010, the male nested with another female in this same nest.

"Now we will see if he has chicks again this year," said Margy as she pulled a can of live mice out of her backpack. While Spotted Owls prefer flying squirrels, they will also accept bushy-tailed woodrats, pikas, and apparently the odd domestic white mouse from biologists. "If he takes this mouse to the female on the nest, they have either eggs in the nest or young. If he eats it himself, they're not on the nest yet. Do you want to hold the stick?"

"What do you mean, 'hold the stick'?" I asked.

"I'm going to put this mouse on the end of this branch for the male to take," she explained.

"Okay," I said, unsure of this procedure and not at all thrilled with the possibility of a mouse running up my sleeve.

It was a needless worry, I soon found out. Almost immediately after Margy handed me the branch and placed the mouse on the end, the male flew in on silent wings while his knife-sharp talons, in just the right position, reached for and grabbed the mouse. It happened so fast I wasn't sure of what I'd just seen, except that this owl was a swift and efficient hunter, clearly accustomed to taking mice.

In awe, I watched the male fly off to deliver his prize to the female. The nest, with a mother Spotted Owl sitting in it, turned out to be in the broom with the leaning tree for a roof. The male gave her the mouse amid a flurry of chirps and squeaks before flying back to its nearby branch.

"Looks like they have chicks," said Stan like a proud grandparent. Spotted Owls do not nest every year, and over the years, he had seen some owls raise their young successfully while others failed. In 1992, he counted 120 Spotted Owls on the eastern slopes of the Cascades. In 2013, the population was down to 21. Here in the Taneum Creek area, there used to be ten nests. This year there was only this one. The hope for Spotted Owls on the eastern slopes of the Cascades rested squarely on this nest with what looked like two tiny downy white chicks.

We fed the male several more mice before our time with them was up. Each mouse he took dutifully to his mate, then flew back to within several feet of us, probably hoping for more mice. My eyes were fixed on the two black eyes of the female that stared from the nest as she kept her babies warm and snug. As we turned to leave, I wished these Spotted Owls good luck and a long life.

Evening approached, and we hiked out of the forest, making our way by following a flowing stream, adding a water element to the already tangled and complex route. I smelled the rich, earthy aroma of the living forest around me as I steadied myself from slipping on wet rocks by touching the muddy riverbanks, grabbing green leaves and branches, and leaning on broken logs, all in various stages—life growing from decay, decay coming to life in the ancient cycle of old-growth forest where the Spotted Owls are integral players in the ancient old-growth play.

When we think of Spotted Owls, we often think of the classic temperate rain forest of the Olympic Peninsula in Washington State. That's where I met Kari Williamson, another wildlife biologist tracking Spotted Owls, one afternoon. We planned to check the progress of the one active nest this year that Kari had in her territory between Forks and Port Angeles. I was looking forward to seeing the differences in habitat between the eastern slopes of the Cascades and the western slopes of the Olympics. And if I was lucky, I would get to see older chicks close to fledging.

I planned to meet Kari in Forks, a logging town best known as the setting of the megahit *Twilight* books and movies. As I drove through town, I remembered the story that Margy told me about the time that one of the local restaurants refused to serve her breakfast because she was working as a hooter, in town to count and help recover the Spotted Owls. She was therefore considered to be against the loggers during the so-called "Spotted Owl war"—a time when bumper stickers with catchy slogans like "Kill a Spotted Owl—Save a Logger" and "I Like Spotted Owl—Fried" were common sights.

The focus of both intense love and hate during the 1980s and '90s, this gentle owl was the umbrella species for the environmental movement's quest to save the last remaining fragments of the unique old-growth ecosystem. Some loggers, afraid of losing their jobs, leveled death threats against scientists trying to count the number of Spotted Owls remaining. The practice of clear-cutting had destroyed 85 to 90 percent of the Spotted Owl's prime habitat, the pristine two-thousand-year-old temperate rain forest of the Pacific Northwest. The irony of the situation was that 90 percent of the logging jobs had already been lost in the previous forty years before the Spotted Owl was listed as "threatened," because of the sheer number of trees that had already been cut. Yet even with all this environmental protection, the Spotted Owl has not recovered.

In an unexpected twist, it is the slightly larger cousin of the Spotted Owl, the Barred Owl, that is now also an obstacle to recovery. "I think the owls are feeling the impact of a broken landscape due to logging, as well as a high level of stress caused by Barred Owls moving into their historic areas, some of the best old-growth around. Spotted Owls really do specialize in the old woods," Kari explained about the Spotted Owl's lack of recovery.

I met Kari at the Forks Transit Center and rode with her to the area where we would begin our hike. "I sometimes feel like I'm documenting the extinction of a species," Kari admitted as we drove to the north end of town. "But I always hope for a good surprise when I go out on long searches looking for more."

On her last check of the nest we would be visiting this day, she had gotten one of those good surprises. There were two healthy chicks in the nest cavity, and Kari had her first look at the mother. It was not the female owl she had expected it would be. It was an unbanded female that had miraculously ended up here after the male's first mate disappeared. The male and his first mate had been together since 2008, nesting successfully in 2010 and trying again unsuccessfully in 2012. Now the new owl pair had raised their two chicks to the edge of fledging. Perhaps today we could count the two chicks as fledged, considered a reproductive success. But first, we had to find them.

Again, I would be hiking straight up a mountain with no switchbacks. Our destination was at an elevation of 1,500 feet in old-growth habitat. We walked first to a shallow river and changed shoes before attempting the slippery ford. This forest seemed more familiar to me than the eastern slopes of the Cascades as it was nearer to my home and I recognized some of the vegetation we encountered along the way: blueberries, salmonberries, sword ferns, witch's hair—a long lime-green lichen that hangs from the branches of trees—downed decaying hemlocks, and moss covering everything. There was more moisture in this forest, but that didn't help my footing at all. I slipped and slid, one step forward and two steps back, up the steep grade.

An hour later, Kari and I made it to the nest site. I stood catching my breath as Kari checked the nest cavity. Empty. The owl family was gone. Had the chicks made it to fledging? Or had something happened to the chicks, causing the parents to abandon the nest? I thought we would never know what had happened, but Kari was not ready to give up the idea of finding this family.

She hooted a Spotted Owl hoot using her own clear voice. We listened in silence for a response. We heard nothing. Then Kari hooted again. This time, a distant *wwoo, wwoo, hooo, hooo* sounded a few hundred yards up the mountain. We started in the direction of the call.

As we scanned the dense forest closely, one owl flew over our heads. “It’s the mother,” Kari said, pointing out the owl’s new band.

Then we saw the father. Kari took off her backpack, set it on the ground, and reached for the coffee can of mice she had packed in. “This should tell us the story of the chicks,” she said.

Holding the mouse by the tail, she placed it on a stump between us and the father owl. With swift focus and agile talons, the Spotted Owl father swooped in and grabbed the mouse. Without hesitation, he took the mouse to the mother, who flew immediately to a leaning branch we hadn’t noticed. There sat two fuzzy off-white chicks as big as the parents.

Food-begging calls rose from the two babies as the mother gave one a mouse. She then turned and chirped to the father, as if asking for another.

Kari put out another mouse, and the father again showed off his swift mouse-catching skills and took the mouse to the mother to feed to the next chick.

Amazed, I watched the proceedings with fascination. Kari grabbed her notebook and, like a Jane Goodall of owls, took notes on each behavior she witnessed over the next two hours that we watched this wild Spotted family.

About half an hour into our observation, we watched as the babies flew after their mother to a branch a bit farther up the hill. They were now good, strong fliers, and Kari could count them as fledged.

Later, one chick held a mouse in its beak as if it was thinking of swallowing the mouse backward instead of the usual way—headfirst. Mom then tried to take the mouse from the chick. But the baby held the mouse tightly, refusing to give up its prize. The chick then proceeded to choke it down tailfirst, the mouse's beady black eyes and whiskers disappearing last down the baby's throat.

The other baby, now sitting in what looked like a cache tree, where the father stored extra food for later, seemed to find the remains of a flying squirrel, which Kari managed to identify by the shape and size of its tail just before the baby swallowed the meal.

Then the father flew to within ten feet of us, sat on a branch, and peered at us through the notch of a tree. I admired his dark-brown feathers decorated with white spots and his kind, dark eyes. Although I knew this medium-size owl was an efficient predator, in this moment he seemed more like a longtime friend.

I tried to take in all the sounds of the owl family and the deep, green scent of the forest so I would remember this exquisite afternoon. I'd had a rare glimpse into the life of a wild Spotted Owl family. I felt full and grateful for the experience.

"Sometimes I'm walking along in a zone or just trying to keep myself alive going down the mountain and I hear an owl I wasn't expecting," Kari said as we made our way back down the mountain. "That's where the magic is. It happens once or twice a season. That's what keeps me going."

We crossed the stream and left the forest, going back to our lives, leaving the Spotted Owls in peace to live their own lives.

Opportunistic

Barred Owls

Laverne sat on a tree branch near a dark road on Bainbridge Island during the middle of a winter night—the same night that the small group of owlers, of which I was a part, came looking for this particular Barred Owl. She had large dark eyes and a sharp yellow beak. Alternating brown and white concentric circles framed her face. Brown-and-white horizontal bars created a hood over her head and a collar around her neck, as if she were wearing a scarf. On her chest she wore the vertical brown bars for which the Barred Owl is named. Her long brown-and-white tail hung below the branch. A bright spotlight in the night illuminated the tree, creating a glow about the striking bird.

What a beautiful owl, I thought.

This was my first interaction with a Barred Owl on my first night of owling, the night I watched Jamie Acker trap and band

Saw-whet Owls on Bainbridge Island. I had heard of the Barred Owls of Bainbridge Island from friends. "They kept me up all night with all their screeching," one friend had told me of the Barred Owl's caterwauling mating calls. "I was coming home late one night and there was this owl sitting right there on the stop sign," another friend said of this common sighting around the island.

Jamie, who had been following the increasing population of Barred Owls on the island, was once attacked by a male Barred Owl guarding his territory near Gazzam Lake. Signs warning of owl attacks began showing up on information boards at park entrances around the island. I had heard the familiar hooting *Who cooks for you. Who cooks the food?* call from the cabin in the woods where I'd once lived. But I had not seen a Barred Owl up close and personal until this owling adventure.

Jamie had banded a number of Barred Owls on Bainbridge, and periodically, he checked on them using domestic white mice, just like the Spotted Owl hooters. Laverne, who seemed to know the sound of Jamie's car, flew to a branch overlooking the road that night when she heard us coming. Then she flew to a closer branch to better watch for her free mouse, giving all of us owlers a great look at this beautiful owl.

Laverne was patient on this dark night as Jamie put the mouse atop a post that looked much like a cat's scratching post. Like the white mouse at the end of the branch I'd held for the Spotted Owl, this mouse stayed put, sniffing and observing its new surroundings, rather helpless in the spotlight Jamie held on it. All I could think about in this brief moment was the scene from the movie *Jurassic Park* where the unsuspecting goat is lowered into the cage of the *Tyrannosaurus rex*. Before the white mouse could make the choice to run, Laverne swooped in and claimed the free meal, as any intelligent predator trying to make a living in the wild would do.

"As individuals I love them," Jamie Acker said of his Barred Owl friends. "But as a species, they're a problem."

Over the last one hundred years, Barred Owls have moved into the Pacific Northwest, and theories abound about how and why Barred Owls emigrated from eastern North America. One theory speculates that it was the result of European settlement.

In the late 1800s, white settlers with dreams of a better life headed west, expanding the boundaries of the nation. As they settled the northern Great Plains, they pushed Native Americans off the land, claiming the territory for farms and homesteads. They also ended the Native American practice of burning the prairies, allowing trees to grow where none had grown before. They planted their own trees and altered the habitat forever. They killed the buffalo and declared war on wolves, removing balance from the landscape. Coyotes filled the hole left by the extermination of wolves, and other animals followed, flowing west with opportunity.

Following the lush green habitat bridge created by this westward movement, possibly as early as the 1920s, Barred Owls—the more aggressive, slightly larger cousin of the Northern Spotted Owl—moved west. Being highly adaptable to a wide variety of habitats in various states of disturbance, Barred Owls thrived in both the pristine old-growth and clear-cut forests. The clear-cut forests may have also inadvertently provided Barred Owls with protection from their fiercest predators, the Northern Goshawks and Great Horned Owls that prefer undisturbed areas for foraging. Eventually, the Barred Owls found an almost empty niche left vacant by the near extinction of the Spotted Owl. The Barreds claimed territories, pushing the remaining native owls out of their preferred habitat, these cool forests teeming with prey.

Other theories suggest that Barred Owls may have moved into the Pacific Northwest from Canada. Did Barred Owls at some point in their evolution overcome some unknown biological hurdle and adapt to the evergreen forests of eastern Canada? Did they make their way west following Canada's boreal forests? They were first documented in British Columbia in 1943 with the first nest site spotted in 1946. They were documented in Western Washington in 1969. By 1979, they had spread into Oregon, and into California by 1985.

Or did Barred Owls spread naturally and normally as part of the species' successful growth and expansion as a predator? Near my home on Bainbridge Island, Barred Owls were first seen in 1992. Little more than twenty years later, in 2014, Jamie Acker monitored thirty-five to forty-five nesting pairs there.

Barred Owls are now considered an invasive species, unwelcome in the forests of the Pacific Northwest. If the pioneer theory is correct about how Barred Owls spread west, it seems that the Northern Spotted Owl's fate may have been sealed as early as the late 1800s, when we did not yet understand the threads that bind species together.

Barred Owls are the opportunists of the owl world. Like coyotes, Glaucous Gulls, rats, and cockroaches, Barred Owls are not picky about what they consume. As medium-size owls slightly over two pounds with a wingspan of three and a half feet, they are efficient hunters. Mice, voles, and shrews are the usual prey on the menu as well as bats, rabbits, opossums, and squirrels, along with smaller owls, such as Northern Saw-whet Owls, Northern Pygmy Owls, Long-eared Owls, and Western Screech Owls.

Jamie Acker saw his last Western Screech Owl on Bainbridge Island in 2009. These small owls hatch and fledge at the same time that Barred Owls are looking for food for their own growing chicks. The young owls, called "branchers" before they fledge, sit on a branch near the nest and cry for food, making them easy targets for Barred Owls. Eastern Screech Owls chicks do not behave in this way because in the east, the two species have evolved together. The Eastern Screech Owls have had time to learn what works for their survival and what does not.

In the Pacific Northwest, we don't have the luxury of time when it comes to threatened species. Barred Owls, the aggressor, are doing their best to claim the old-growth territory from the threatened Northern Spotted Owl. And therein lies the problem.

Dave Wiens, a wildlife biologist with the USGS in Corvallis, Oregon, studied the competition between Northern Spotted Owls and Barred Owls for his PhD dissertation at Oregon State University. He found a high level of competition between the two species, with Barred Owl pairs outnumbering the less aggressive Northern Spotted Owl pairs by four to one. He also estimated that Northern Spotted Owls require about two to four times the area per home range of Barred Owls—4,554 acres to 1,435 acres, or seven-square miles to 2.24-square miles, respectively.

He then radio-tagged twenty-nine Spotted Owls and twenty-eight Barred Owls in a 463-square-mile study area and followed them for two years, from 2007 to 2009. He studied how these species use their territory, what type of habitats each species chooses, what they eat, and survival and reproductive rates.

The diet of Barred and Spotted Owls is similar in that both prefer flying squirrels, woodrats, and hares. These tasty species that account for 81 percent of a Northern Spotted Owl's diet only take up 49 percent of a Barred's menu. One day during his study, while observing a radio-tagged female Barred Owl sitting in a tree above a shallow pool, Dave saw the owl drop down to catch crayfish, one after another. She had no need to move beyond the tree at all. He watched other pairs of Barreds doing the same thing. They also caught insects, salamanders, and frogs, prey that also did not evolve alongside this nocturnal avian predator. Barred Owls, it seems, are outcompeting many species in the Pacific Northwest.

Yet in all the time Dave spent watching the Barred Owls during his two-year study, he couldn't help but get to know the individual personalities of the owls. "I have a huge amount of respect for both species, especially the Barred Owls," he said. "They're an impressive species. Very intelligent and adaptive."

However they arrived, they are now members of the web of life in the Pacific Northwest. It may matter only to us humans as we try to save the Spotted Owls from certain extinction. If we label the Barred Owls as invasive, as interlopers who do not belong in our woods, it becomes easier to remove them. If they have arrived naturally as nature evolves and expands, then removing them becomes a tangled and complex moral dilemma.

My own experience with Barred Owls, since that first night with Laverne, has been among the more personal of my owl journey. It was the Barred Owl that first captured my attention at the beginning of my exploration. While at a coffee shop, meeting a friend, I walked past an exhibit called *Owl People* by photographer Mark McKnight, who lives in the same town as I do. He had photographed the Barred Owls in his yard, and the intimate images of owls sitting on branches draped with lichen in the mists of the Northwest forests instantly captured my imagination. I stopped what I was doing to stare for a moment into the dark, round photographed eyes of the beautiful Barred Owls. I was deeply moved by the brief experience, and after that, I noticed images of owls everywhere. With all this synchronicity, it seemed as if the owls had a message to share with me.

A few months later, a family of Barred Owls flew into my neighborhood. I first noticed them from their raucous calls in a nearby tree. *Who cooks for you. Who cooks the food?* I heard, and I ran outside to see two owls fly from the tall Douglas fir next door, across yards to the next street over. I was thrilled to have owls so near my home. I stood listening as the last of the daylight disappeared and wondered if anyone else had noticed them.

Who cooks for you. Who cooks the fooood? called the first owl, on the right and closer to my ears. Like an owl, I was learning to use my ears to pinpoint location.

Who cooks the fooood? the other, more distant owl responded.

They called for almost an hour. I tried to guess at their conversation. Barred owls mate for as long as they survive, but this didn't sound like two adults to me. It sounded like one of the owls was begging for food, so I guessed that one was a chick refining her hunting skills. The pair flew to another yard across the hill. As the stars brightened, the two Barred Owls continued to talk. I heard them in the same Douglas fir for the next several nights, calling at dusk. Each time I listened.

On another night about two weeks later, my neighbor Ann Bernheisel called in a panic to tell me she had just seen a bear. When she heard a noise, she stuck her head out of her bedroom window to investigate. And there, standing on her shed, trying to steal the birdseed hanging between her window and the small structure, stood a black bear. Nearly nose to nose with

Ann, the frightened bear slid off the shed and onto the ground with a crash. Hoping to see the bear, I ran outside to meet Ann at the scene of the attempted crime. As we stood in the dark, all of our senses were on high alert, and we only had dim flashlights to show the way. A scream pierced the stillness.

"That's human," whispered Ann.

"No, that's an owl," I whispered back, heart racing.

The owl called again. *Whoooo!*

"That's an alarm call," I said, in my mind seeing the frightened bear climbing the startled owl's tree.

Over the next month, I heard nothing from the owls and figured they had gone. Where they would have gone, I had no idea, since Barred Owls stay on their home territory year-round. Then, on the night of the Geminid meteor shower in mid-December, as I lay in my down sleeping bag watching the sky and listening to the live-streamed whistle of falling meteors on the Slooh website, I heard a Barred Owl in a tree across the road. *Whoooo,* the owl called with the telltale guttural gurgle of the Barred. The owl called again, and I listened as long as he spoke.

I was falling in love with my feathered neighbors. But I was torn, because they didn't belong here. I knew what they were doing to the Spotted Owls, the Saw-whet Owls, the Western Screech Owls, and the Pygmy Owls: chasing them off to claim territory for themselves and making meals out of smaller owls, who had not yet adapted to their presence. But these Barred Owls were becoming my friends, my companions, by proximity and familiarity.

Now we in the Pacific Northwest are faced with a complicated dilemma. Some biologists have proposed the systematic removal of the invasive Barred Owl from the land we have set aside for the recovery of the Northern Spotted Owl. Currently, the US Forest Service is conducting a five-year pilot study to determine if removing the Barred Owls from this territory will make a difference in the recovery of the Spotted Owl. In September and October 2015 , sharpshooters killed seventy Barred Owls in an experimental territory near Cle Elum. In another separate territory, designated as the control area for the study, Barred Owls remain undisturbed. Upon completion in 2020, the results will be analyzed to see if the Northern Spotted Owl benefited from removal of the Barred Owls. The Forest Service will take public comments, and a determination will be made about how to proceed. This study will provide the framework for future management decisions.

Some lucky people remember the call of the gentle Spotted Owl while camping or hiking in the old-growth forests of the Olympics and the eastern slopes of the North Cascades. Those people understand on a primal level that it would be tragic to lose this species forever from our forests—the very forests that help to provide the oxygen we breathe. But because humans may have played a role, however unwittingly, in the demise of the Spotted Owl by destroying its forest homes, we have a responsibility for the owl's recovery.

Spotted Owls have an important role, helping to sustain the forests by doing what they were designed to do, playing their part in the scenes of forest predator and prey. What will happen to the forest cycle if it loses one of its supporting players?

Human actions, as unconscious and unintentional as they were at the time, may have caused the Barred Owls to follow us out west. The irony of the Spotted Owl/Barred Owl situation is not lost on me, as a white person living on land held by the Suquamish Tribe. I listen to and love these Barred invaders in the homeland of Chief Sealth, better known as Chief Seattle, who famously called for all humans to remember to live with the land, because each strand in the tapestry of life is as important as the next to the whole.

Only now, in these early years of the twenty-first century, are we beginning to understand through careful study of natural systems how these threads in the tapestry are interrelated. Miraculous ways, like how the nutrients in decaying salmon, scattered throughout the forest by the animals that eat the fish, are as important to the trees as the rain that blesses our Northwest forests. Because we are studying the broken remains of a once-stable system thousands of years old, now, in an ever-changing state of flux, we can never know how many strands are safe to pull or remove altogether.

Some have intuited an answer from their own thorough observations of what they noticed around them. Nearly sixty years after Chief Seattle gave his historic speech, the author and conservationist, Aldo Leopold, shared his thoughts in his essay "Thinking Like a Mountain." Leopold warns of removing one species in favor of another. "The cowman who cleans his range of wolves does not realize that he is taking over the wolf's job of trimming the herd to fit the range. He has not learned to think like a mountain. Hence we have dustbowls, and rivers washing the future into the sea."

I do not have a good answer or another solution for this agonizing dilemma. But I do wonder, what if one day we realize we needed the Barred Owl to help us rejuvenate the old-growth rain forests or some other job that we cannot yet see? What will we do if we have removed the wrong thread?

Underground

Burrowing Owls

The golden landscape felt foreign to me as I drove diagonally across southeastern Washington on I-82, through Yakima and the Tri-Cities, crossing the Columbia River into northeast Oregon. Coming from Western Washington, defined by Puget Sound and its lush green shores lined with tall Douglas firs, western red cedars, and curvy red-barked madrones, I felt as if I'd driven into another world. On the other side of the Cascades were vast skies, tall sloping foothills, and wide-open grasslands—Burrowing Owl territory.

Before I saw any owls, I saw the "igloos"—man-made grass-covered concrete bunkers built to store chemical weapons, like mustard gas and other nerve agents, from a bygone era, when we first thought to poison people on the other side of the globe. Now the igloos stood empty on the seventeen thousand acres of the US Army's Umatilla Chemical Depot, a few miles from the Washington-Oregon border. As a result of the Chemical Weapons Convention of 1993, which banned the production, use, and stockpiling of chemical weapons, the supply stored here was incinerated by 2012.

Now, as the base has become a burden to the US Army, it has become an oasis for others. Burrowing Owls—with their white Groucho Marx eyebrows, white spots on their brown foreheads and wings, and skinny bare legs—come here to breed and raise their young. On the smaller end of the owl spectrum, about the size of a Steller's Jay, the Burrowing Owl displays a meerkat-like charisma.

I had come here to meet these unique underground owls and to meet owl savant David H. Johnson, who has studied owls ever since he met his first owl at twelve years old, while on a camping trip in Minnesota, where he grew up. "I could see the silhouette of the Screech Owl on my tent," David told me. "It sat there for twenty minutes, calling. It could've chosen to sit anywhere, but it chose my tent. I decided right there to devote my life to studying owls."

It's not every day someone answers an owl's call by choosing to make a living learning about owls. But not only has David spent a lifetime apprenticing himself to owl ways and looking at the place of owls in culture, he has put his life on the line at times. So far during his career, he has fought for the Spotted Owl in the Pacific Northwest, defining the legal, critical habitat lines around Spotted Owl territories and testifying in court about the biology of the threatened owls, all while receiving death threats as thanks for his work. He's counted the calls of Saw-whet and Boreal Owls in the dead of a Minnesota winter, captured Flammulated Owls in Washington and California, and driven around the western United States through Oregon, Washington, Idaho, Montana, and the Dakota Badlands, capturing and banding Burrowing Owls. He helps coordinate graduate student projects, organizes symposiums, and, in his spare time, carves Pacific Northwest–style owl masks.

He has also conducted a worldwide study on beliefs and cultural attitudes toward owls in hopes of understanding how opinions impact owl conservation in thirty countries around the world. "We used to kill witches," David said. "Now we dress up like them. We have taken what was considered dangerous or significantly powerful and turned it around. They no longer hold power over us." His question, he explained, is "How do we do that with owls?" The results of his study will be published in a special edition of *Journal of Ethnobiology*, or "the interdisciplinary study of the relationships of plants and animals with human cultures worldwide."

Because owls live on all continents except Antarctica, they are powerful symbols for conservation, and as umbrella species their presence could have lasting effects in conserving the land on which they live. In 2017, the Smithsonian National Museum of Natural History will showcase an exhibition of owl artifacts and stories collected by David entitled *Spirit Wings: Owls in Myth and*

Culture, highlighting the owl's global association with human culture over millennia.

When I met him at the Umatilla Chemical Depot, he was in the process of building a special photographic burrow to be used by a National Geographic photographer for a story on Burrowing Owls. He is a walking encyclopedia of the world's living and breathing owls. Even his circadian rhythms have shifted to more closely resemble those of an owl. On a couple of nights, we sat in his office talking owls to well after midnight until my brain could hold no more.

He founded the Global Owl Project in 2002, he told me, after consulting on a project for NASA about how to catalog life on any new planets that may be discovered. But as interesting as life on other planets may be, David Johnson had his mind set on planet Earth. "Never mind other worlds, we don't even know what we have on this planet," he said. His goal with GLOW is to learn about all 642 of the world's owl species and subspecies.

Another of his current projects is directed toward helping to reestablish the diminishing population of Burrowing Owls in the Pacific Northwest. "It's the best show in the West," David said of his project on "the Depot" during our first phone conversation. "In 2008, when we started, we had only two to three pairs on the Depot. In 2009, we had six to eight. Now, we have thirty-one pairs." And he invited me to come and see for myself.

I arrived at the Depot in the spring of 2014 as a first-time volunteer. Once I was cleared by the guards to enter the army base, David met me outside the firehouse, designated base camp for this year's project. "Ready to go?" he asked. I nodded. "Let's go check some burrows."

I quickly stashed my luggage in a room on the second floor, threw my binoculars, a notebook, water, and a snack into my backpack, and grabbed my camera on the way out. Signs of over seventy years of army occupation were visible on the base in the form of rows of vacant igloos, decaying munitions buildings with evenly stacked red bricks designed to fall in on themselves in case of explosion, empty storage units with rusting equipment exposed to the elements, and small square buildings with broken windows that now serve as shelter for Barn Owl and raven nests far away from the speeding vehicles on nearby country roads.

Fast-moving cars and trucks represent one of the greatest threats to all owls. (The other man-made threat is secondary poisoning from rodenticides.) I'd once found a Barn Owl dead by the side of the road as I drove through the wide-open fields and croplands south of Yakima. Without a scratch on its tawny body flecked with black specks or a cut on its white face with beautiful owl eyes wide open, the owl must have been killed instantly by a passing car. As the light had grown dim and the owl had flown across the road focused on her next meal, she was unaware of the oncoming danger.

Now that all the chemical weapons have been incinerated, with air, water, and soil samples taken 24–7 to ensure the safety of the surrounding landscape, the army plans to hand over 5,500 acres as a wildlife preserve with Burrowing Owls as the poster species for conservation. David hopes to establish a Friends of the Depot organization to help oversee the recovery of this charismatic species. And all the owls on the Depot will be able to hunt in relative peace, free from human-induced threats with only natural predators to avoid.

The Columbia River, winding its way toward the Pacific Ocean, shone in the distance as we arrived at the north end of the base. It was uncommonly quiet here. Distant calls from nesting curlews, temporary residents from the ocean's distant shores, broke the silence. David pulled out the map to show me how the owl burrows are arranged.

Burrowing Owls position themselves in loose colonies, which seem to have some definite benefits. Males count on one another to keep an eye on the horizon for danger. If one owl sees a coyote or a hawk, he sounds an alarm call for all to hear, giving each family a chance to run for the cover of the burrow. It may also be easier to find a mate in a colony rather than at a burrow that is far from its nearest neighbor.

Just as other owls don't build their own nests, most Burrowing Owls do not dig their own burrows. The exception is a subspecies of resident Burrowing Owls found in Florida, which will dig their own burrows with as little encouragement as a shovelful of dirt dug out of the front yard of a home. In the western United States, Burrowing Owls choose for their homes abandoned prairie dog or badger holes. That's a problem for the owls in Washington and Oregon, however, because the badgers that primarily dig the burrows here are disappearing. Many badgers on the Depot were caught and killed as bycatch in traps set for coyote removal. Also badgers are considered by farmers in the area to be a nuisance animal and are often shot. Cattle grazing in the fields can break a leg if they step wrong into a burrow.

Because reintroducing badgers to the area is not a popular solution among the human residents, David has placed eighty-six artificial burrows on the Depot in the same loose colony arrangement found in nature, providing housing for the Burrowing Owls until the badgers can return naturally to the depot. (David estimates that the Depot has a food supply for as many as thirty badgers.) Each man-made burrow is constructed using one half of a fifty-five-gallon barrel with a tunnel made of ten feet of black PVC pipe attached on one side of the barrel. A hole is cut in the barrel's top, and a five-gallon bucket without the bottom is glued over the hole in the barrel, to create an easy entryway for checking burrows. Then a second five-gallon bucket with a lid is stacked on top of the first bucket, to close the top of the burrow.

After digging a large hole for a barrel burrow and its ten-foot tunnel, David places a screen of wire mesh beneath each barrel to keep pocket gophers from filling the burrow with sandy soil. Pocket gophers apparently do not like to share their area and will fill any empty hole with sand.

Finally, the burrows are buried. When it's time to check the burrows for owl occupancy or to band the adults or babies, the tunnel is blocked so no owls can escape and the bucket on top is lifted out for an easy capture.

Each Burrowing Owl site, chosen and guarded by one male, includes two burrows about ten feet apart and a "T perch" made of two wooden two-by-fours nailed in "T" shape for the males to sit on and survey the surrounding steppe for predators and prey. The sites are placed about a quarter mile apart. One of the two burrows in each site is used by the female to lay her eggs and raise her young. The male uses the other burrow as his "man cave." The proud male often caches food there, like kangaroo rats, mice, and pocket gophers, whole or beheaded, stored neatly with the headless end against the wall of the burrow to keep sand out of the decapitated body. During one very good year for Burrowing Owls, one male, hunting during the crepuscular hours of dawn and dusk, managed to cache fourteen small mammals, setting a new Depot record.

Male Burrowing Owls also "decorate"—with their own exacting "taste." Scattered around the burrow's entrance are items such as corn husks or cobs, coyote and cow dung, bits of cryptogamic crust that look like little pieces of moss, parts of prickly pear cacti, and even short strips of rubber tire tubing. Each owl has his own style, showing any potential interlopers that the territory has been claimed.

Some scientists believe that the gathering of dung may be for more than just territory marking or decoration. In an article published in September 2004 in the journal *Nature*, Douglas Levey, Scot Duncan, and Carrie Levins suggested that it's also a type of tool use, that the owls are gathering mammalian dung as a means of "fishing" for dung beetles.

The trio tested the theory on the population of Burrowing Owls in Florida. After removing expelled owl pellets, dung, and beetle parts from all the burrows they were observing, the biologists placed about two hundred grams of cow dung (about a half pound, the usual amount collected by Burrowing Owls on their own) at the entrance to the burrows in one area, creating the experimental group, while the other burrows in the control group got no dung. After four days, pellets and prey remains were collected and counted. Then they reversed the two groups, giving the control group their share of dung this time while the other group got none.

The research team found that owls in the burrows with dung ate ten times more beetles than the owls without the dung—a significant finding substantiating the use of dung as a tool by Burrowing Owls to catch large black dung beetles.

In the Pacific Northwest, there are four hundred species of darkling beetles for Burrowing Owls to feed on. From time to time, they also enjoy a tasty red Jerusalem cricket, june bugs, large bees, deer mice, kangaroo rats, and small gopher snakes, running as much as flying to catch their meals.

On this day at the Depot in late April, David and I began the process of checking each burrow for occupancy. By this time of year, many male owls had arrived, claimed territories, and attracted a female with the tempting offer of a mouse or kangaroo rat. Already some of the first clutches were being laid.

We drove to the first burrow, removing tall green tumble mustard, cheat grass, and other vegetation from the sandy patch in front of the burrow entrance, and from the area on top of and behind the burrow, to create a clear line of sight for the owls to easily spot predators. Next, to prevent the mother from flying away, David rolled up his T-shirt to block the entrance to the burrow he suspected of being the nest burrow. The nest burrow was more likely than the other to have small white feathers outside the entrance, shed from the female's brood patch—the bare red swollen spot just above the mother's legs from her belly to her chest, which gives maximum body warmth to developing eggs and chicks as she sits and broods.

"Let's take a look," David said as he lifted the bucket off the top of the burrow. On hands and knees, I peered into the darkness, careful to avoid any black widow spiders with the distinctive red hourglass shape on the belly lurking within. When my eyes adjusted from the bright sun to the dim burrow, I could see round quarter-size white eggs, lain directly on the dirt. The mother stood on guard, looking up at us with bright yellow eyes.

Startled by the light, one downy white chick wriggled between the eggs, with the bulging blue eyes of the very new shut tight. Another chick, engaged in the pushing, wriggling struggle of birth, fought to break free from the half-on, half-off eggshell that had, until this moment, held its tiny life.

David counted quickly. "Seven, eight, nine eggs and two chicks," he said as a proud smile swept across his face. "That's a good start, eh?"

I nodded in silent agreement, stunned by the honor of witnessing this new life.

We would spend the next two days checking and cleaning burrows. When we came to a less-popular-with-the-owls area of the Depot, we checked and cleaned anyway. "Not all of the nests will make it," David said. "There may not be enough food. The male or female may die, and they can't do it alone. If there is enough time, they may find another mate and another burrow to start over. It's all about reproducing. They have no choice. They must go on."

It was midafternoon by now and I was leaning over my sixth empty burrow, sunburned and thirsty, while weeding. *They must go on* kept repeating in my head. *I will do this for the owls*, I thought with each weed I pulled. I thought about the amount of energy given by each parent. The female Burrowing Owl will lose 33 percent of her 220 grams laying eggs and raising young. She is dependent on her chosen partner for her food. For sixty days while she is on the nest, caring for her brood, she will leave the burrow for only short breaks. If the male is a good provider, she and her chicks will thrive. If not, they may perish. If he doesn't come back, for whatever reason, she will have to make the choice between the lives of her chicks or her own life. She will not stay. She may then find another mate, putting her life at risk again, trying to raise a family when she is less healthy with fewer resources available to her than she had at the beginning of the nesting season.

Life in the wild, in this desert landscape, is perilous. There are no guarantees. Some survive and live to pass on their genes to the next generation. Others do not. As we worked, David told the story of one female Burrowing Owl who, under severe conditions, lost all eight of her chicks. Whether she killed them herself or they were already dead from starvation is unclear. Yet the mother, in her desperate will to survive, ate her own chicks, taking their bodies into hers in a sacred cycle of life and precious resources. Nature does not worship death. She gives all her resources to the living and is unflinching in her never-ending quest for life. The mother recovered, and when conditions improved, she laid another clutch with the same male. Together the pair successfully raised five young in the same season.

"It's really quite amazing," David said. "To be so close to death . . . and come back that far. That is the kind of thing that drives evolution!"

They must go on.

Evolutionary time is not always on the side of the Burrowing Owl. As global climate change sets in, one day of significant rainfall can cause flash flooding that can drown the owls nesting in their underground burrows. Or only four days of continuous cold, rainy weather—not uncommon for the Pacific Northwest in the spring—at the wrong time in the nesting cycle can cause owl young to starve because their fathers can't hunt, wiping out an entire generation.

Changes in prey abundance can also cause shifts in the Burrowing Owl population. In 2011, when the number of nesting pairs was high, at sixty-one pairs, David's team banded 203 chicks at twenty to twenty-four days old—the standard measure of reproductive success for Burrowing Owls on the Depot. The next year, 2012, there were sixty-five nesting pairs on the Depot—four more pairs than the year before. Yet only 103 banding-age chicks were banded, half of the number from the previous year, after a catastrophic midseason collapse of prey species. Was it climate change, natural cyclical changes in small mammals, a decline in food resources, or some other variable that caused the collapse?

"I have no idea why," said David. "I didn't know the owl populations fluctuated like that. We're learning a lot."

The year 2013 was also dismal for nesting, with only thirty-nine pairs successfully raising sixty-three chicks. "Did the owls say, 'Last year was such a bust I'm not going back there again'?" said David.

It's hard to know, but in 2014, their numbers began to rebound, with thirty-one nesting pairs rearing 163 banding-age chicks.

The next nesting season was an even better year for the Burrowing Owls. There were fifty-three nesting pairs confirmed on the Depot, with some owls that had been banded as chicks returning as adults to raise their own families there. On

June 29, 2015, David posted a note on the Global Owl Project's Facebook page announcing "As of last night we have banded (exactly) 200 young. Most impressive!!! There are still 11 young in natural burrows that we are trying to catch for banding. Also there are 2 females who lost their first broods and are re-nesting. One is incubating 5 eggs, the other 3 eggs." It was the most successful year to date for Burrowing Owls on the Depot.

Burrowing Owls are not endangered now, but numerous small decisions—conservation laws, public opinion about the presence of badgers that provide burrows for the owls, and the placement of a shopping mall, housing development, or paved parking lot—can have a major impact on these little owls and how they are able to survive. No matter how resourceful they are, some of what it takes for them to survive is out of their control.

But it is within ours. We get to choose how we live on the land.

After several days of cleaning burrow sites, it was time to begin trapping and banding all of the adults, whether they were nesting or single and looking for a mate. David assembled a team of volunteers to help with this tricky task. There was Jessica Giordano from Seattle, who had fallen in love with owls during her volunteer vacation at the Depot the year before. Claire Tahon from Belgium, here as an assistant for the summer to learn all she could about a variety of owls in the western United States. And Holly McLean, a graduate student from nearby Richland, Washington, working on her master's thesis on Burrowing Owls. Together we worked to add important information about the Burrowing Owl population on the Depot.

One afternoon, I went with Holly, Jessica, and Claire to finish a group of burrows on the western edge of the Depot. Each section we finished felt like a milestone, but I was particularly excited to be going back to this nest. It was the same nest David and I had checked earlier in the season with the mother on nine eggs and two chicks. I was curious to see the new hatchlings.

We caught each adult either in the burrow or in a wire trap built by David, much like a Havahart trap, in which mice and other small animals are caught but not killed. The female Burrowing Owls were usually the easiest to catch. David had duct-taped a towel the width of the tunnel, four to six inches, to a ten-foot-long section of PVC pipe to create a plunger. Jessica slowly pushed the plunger into the tunnel and corralled the female in the nesting cavity of the burrow. I found the bucket's handle buried in the dirt on top of the burrow and lifted the bucket off the nest chamber. Then, hunched over the bucket-size opening, Jessica and I peered into the darkness. A mass of wiggling fuzzy white baby owls slowly appeared. Quickly Jessica counted. I was having trouble distinguishing each chick from its siblings and wasn't sure I had counted them all. "Eleven chicks," she said confidently. All the eggs had hatched!

"That's great!" said Claire, writing the number on the data sheet. Then Jessica reached in and, with one hand, grabbed the mother huddled against the wall of the burrow and held the mother's wings against her body.

If an owl was new to the Depot, we gave it its first leg band, but this owl had nested on the Depot before and already had one. So Jessica read the number on the band—1004-11414—to Claire, who wrote it down on the data page.

Then Holly took the mother in her lap and carefully arranged her. She stretched out the little owl's right wing and measured her wing from the bend to the tip of her longest flight feather. I could see white bars across the delicate brown wing. Recent research on the chemical composition of owl feathers has shown that not only are these white bars decorative, they also provide the "skin" of the wing while the brown part of the feather provides the structure—like the form of the wing on the first plane built by the Wright brothers in 1903.

"Twenty point zero centimeters," said Holly, reading the digital caliper on the small owl with an almost two-foot wingspan.

Next, Holly measured the female Burrowing Owl's tarsus, the leg from the heel that looks more like a backward knee to the toes. Then, gently, she measured the length of the owl's tail from where the central tail feather connected at the body to its tip. I watched Holly's expert handling of the owl with a sense of awe, the way I'd watched women change a baby's diaper before I had a child. The owl did not struggle but went along with the proceedings, as if she knew trying to escape would be futile. That's a good sign for an owl handler.

Holly also turned the owl over to check for a brood patch. Blowing on the owl's belly and ruffling her feathers would expose the brood patch, if there was one. This female had a warm, swollen, and bare brood patch, confirming that she was currently on the nest. The brood patch of a female with fledged chicks or an unsuccessful nest is dry and flaky. Eventually, when brooding is over, the mother's feathers will grow back in that spot.

Finally, Jessica prepared the scale by placing a plastic Tupperware container and a white cotton bag on the scale and zeroing it out for a baseline. Holly placed the female owl in the bag and pulled the bag closed. Then she laid the little owl in the container, and Jessica read the scale. Time seemed to slow as we waited for the digital scale to settle. Quiet surrounded us on the open grass steppe. We sat in a circle around the small scale with the owl, and for an instant, I felt the sacredness of this circle of women as if we were a necessary part of nature's plan for the recovery of a species.

"One hundred sixty grams," Jessica said, breaking the spell.

It was my turn to release the owl, and I gratefully retrieved her from the bag. I held her as David had taught me, my right hand over the little owl's back, securing her wings, and my left hand holding her legs between my thumb and hand. As I walked with the owl back to her burrow, I could feel her warm body beneath her soft feathers. She was light, her 160 grams almost nothing in my hands. I looked into her yellow eyes as she gazed back at me. "It's okay," I said, intending to reassure the mother that she would soon be back safe with her babies.

I wanted to take my time and hold this wild owl for a while longer. Yet another part of me could feel her fear. What must have this experience been like for her, trapped and taken from her babies the way she was? I crouched down in front of the burrow entrance and positioned the owl to go in. When I released her, she ran to the safety of the cool, dark burrow without looking back. "Good luck, Mama," I whispered, wishing this Burrowing Owl a long, healthy life.

Now it was time to catch the male, which could be a tricky business. But David, thinking like a Burrowing Owl, had already come up with an ingenious new method of trapping the male owls. Previously, biologists used nets with mice as bait. But that often required long hours of waiting for the owls to take the bait.

So David asked himself, *What is the most important thing to a male Burrowing Owl?* The answer was his territory. The males go to great lengths to choose and protect their burrows. A territory with an ample food supply, with two to three burrows on each site, with a perch on a slight hill with a clear line of vision to scan for predators and prey is valuable real estate in Burrowing Owl society. A male with a desirable territory will fight any interloper who tries to come in and take over.

It was this bit of knowledge that David used to devise a scheme to catch the males. He covered a six-inch-wide wire trap with black gardening cloth and placed it in front of the tunnel opening, effectively extending the tunnel by about two feet. Then he placed a speaker inside the tunnel to make it sound as if another male were inside the burrow trying to claim his territory.

It worked like a charm. The strong, dominant males, capable of claiming and holding a territory due to their size, rushed right in to throw the interloper out and were caught quickly. But the less dominant males, the younger or smaller owls, accustomed to being wilier to claim and hold a territory, took a little more time to catch.

That afternoon, Holly, Jessica, Claire, and I used David's trick to catch the male at the burrow where we were working. We carefully removed the decorations the male had placed in front of his "man cave" burrow, where he had stored his cache. Then Jessica and I smoothed out the sand in front of the tunnel entrance to make way for the trap. Holly turned on the MP3 player and started the recorded Burrowing Owl call. We camouflaged the trap with soil and sod from nearby, replaced the decorations, and the trap was set. Now all we had to do was wait for the male to take the bait.

Because Burrowing Owls are wary around humans, they won't come to investigate the burrow if they know humans are close. So we moved on to the next burrow, to trap, measure, and weigh another female. We would come back to check the male's trap in forty-five minutes or an hour. Such was our Burrowing Owl trapping routine.

Curious about how Burrowing Owls reacted to the false rival's call, one volunteer at the Depot set up a remote video camera to film several of the encounters between a male and his false rival. I watched a short video clip with David that night in his office. From outside the camera frame, the male flew in, landing close to the tunnel entrance. *Err-errrr!* called the male, sounding more like a rooster than an owl. Puffed up like a turkey, he strutted around in front of the burrow entrance, acting as if he could not believe the nerve of this intruder. It was hard not to laugh as the angry male confronted the whistling *Err-errr!* of the MP3 player.

As funny as this behavior may be to us, it was life-and-death serious to this little owl. His white eyebrows seemed to frown fiercely as his collar of black feathers stood up around his white-feathered neck. He ran back and forth as if challenging his rival to come out and show himself. When no other owl showed its face, he gathered all his courage and went in after him. And he was trapped, listening to a rival calling, until someone could get back to measure and release him.

After catching, measuring, and weighing another two females and setting two more MP3 traps, we returned to the original trap about an hour later.

Then Jessica crouched and peered into the dark trap. "We got him!" she whispered.

"Yay!" we all cried, a bit tired, hungry, and hot by this point. Yet all our hard work had been worth it for the important data we were gathering about Burrowing Owls.

Jessica reached into the trap and retrieved the mad-as-hell male. Holly weighed and measured him as Claire took data.

[illegible] wls show high site fidelity, with some of the most [illegible] rds returning to their most favored sites year after [illegible] was no surprise that this owl had also been to the [illegible] re; he had a band on his left leg.

[illegible] recent migration study, David tagged Burrowing [illegible] und the West and tracked them by satellite. He [illegible] hat the females traveled the farthest each year. [illegible] les traveled short distances, if at all. For a female owl, [illegible] e goal of migration is to recover from raising her young so that she will be in good enough physical condition to lay eggs again the next year. The female owls tagged on the Depot flew to various locales throughout California for the winter, with one female ending up in the Channel Islands. On the return trip, another female made the seven-hundred-mile journey from Selma, California, to Baker City, Oregon, in seven nights—a Burrowing Owl feat never before recorded. Other female Burrowing Owls tagged in Montana, South Dakota, and Alberta and Saskatchewan, Canada, flew all the way to Mexico for the winter, following the easiest bird migration route down the eastern Rocky Mountains. One owl in western Nebraska traveled to a little town in north-central Mexico in the state of Zacatecas. The town's name, Tecolotes, means "owls" when translated literally.

For male Burrowing Owls, migration is another story. As the providers of the family, they are all about staying close to the best territory. Of the fourteen males tagged at the Depot for the study, four went to California while the other ten actually went north to the Grand Coulee area, to central Eastern Washington near Moses Lake, and still farther north into Canada, in search of a population of small mammals to hunt. The males will risk a harsh winter in order to stay close to their known good territories. For without prey for him to provide his family, a male Burrowing Owl has nothing.

Holly now handed the male to me to release back to his burrow. With my right hand holding his wings secure, my left hand supported his weight and my thumb secured his feet. I stared into the yellow eyes of this Burrowing Owl as I walked him back to his man cave for release. As I had done with his mate, I leaned over, pointed him toward the entrance, and let go.

To my surprise, he did not run immediately inside as the female had done. Instead, he stopped, turned around, and glared at me, as if in challenge, for a few seconds. Then he disappeared into the cool darkness of his burrow.

I laughed out loud at the nerve of this little owl. I admired his fierce will to not only survive but to protect his family through intimidation of a much larger opponent.

In June, I returned to the Depot for a day to help with the banding of the chicks. While Holly put on leg bands, I sat cuddling two of the five almost-adult-but-still-fuzzy babies as they waited for their turn. My mothering instinct kicked in full force as I imagined how much fun it would be to raise a Burrowing Owl. Then good sense returned, and I released the two babies back to their mother and the wild where they belonged. I wished them my now-customary "Good luck," hoping for these little Burrowing Owls' safe returns to the Depot next year—a symbol of a species' recovery.

Synchronicity

Northern Pygmy Owls

The tiny Northern Pygmy Owl (*Glaucidium gnoma*), the second-smallest owl in the Pacific Northwest, defies much of what is considered common knowledge about owls. These owls hunt songbirds that are close to their own size and live their lives during daylight hours. The satellite-dish-like owl face, common to other owls, is ill defined. And Pygmy Owls do not share the asynchronous hatching behavior common to other owls.

I was so intrigued by these mysterious little owls that I tried to see one on several different occasions. But my efforts were to no avail, even in Portland, Oregon, where I looked for the famous population of Northern Pygmy Owls in Forest Park and Tryon Creek State Natural Area. I had almost given up hope of ever finding one.

Then, one night at a party, I ran into my friend Roger van Gelder, a nature photographer. "I saw a NOPO," he said.

"What's that?" I asked, searching the back of my mind for this vaguely familiar-sounding species.

"A Northern Pygmy Owl," he said with a broad smile across his face.

"Oh, I've been looking for one of those! Where did you see it?"

"Chinook Bend Natural Area in Carnation. Do you follow Tweeters?"

I knew about the Tweeters e-mail list where local birders posted recent sightings of different bird species, but I had fallen behind recently in checking it.

"I haven't looked at it in a few weeks," I admitted.

"I went on Monday. It's been there for a few days."

"Did you get some good photos?" I asked, assuming he would have captured the moment in one of his beautifully crisp photographs.

He nodded, grinning.

"I've got to get out there," I said, wheels already turning, checking my calendar for an open date, which would not be for another two days. We made a plan to catch an early morning ferry on the following Saturday.

I awoke early on Saturday morning, dressed in warm layers, and wondered if this trip would end up like the others, all for naught. I didn't know if the owls—there had been two reported—would still be in the area nearly a week after Roger had seen them. The NOPOs hadn't been listed on Tweeters in several days.

There were sightings in other areas, twelve Pygmy Owls reported in the Okanogan and one more in nearby Enumclaw, but had the owls in Carnation flown away?

I thought about canceling the trip and going back to bed on this cold winter morning. But my then-eleven-year-old daughter, Ellie, repeated my own advice back to me. "You won't see anything if you don't go look." So I gathered my gear and went to meet Roger.

When we arrived at Chinook Bend Natural Area, the small parking lot was already quite full. As a wildlife photographer, Roger had brought a heavy 600mm lens, a high-speed digital camera, a tripod, and a spotting scope. I had borrowed another of Roger's lenses for my camera. I had never tried a superlong lens and was curious what kind of photographs I could take with one, but as a naturalist, what I really wanted was a chance to observe the behavior of this unique little owl.

At six to seven inches long with a wingspan near fifteen inches, the Pygmy Owl is a fierce diurnal, or daytime, hunter of a variety of songbirds, including wrens, hummingbirds, warblers, and jays, and small mammals like voles. It also preys on some animals as large as or larger than itself, like red squirrels and California Quail.

Pygmy Owls are known to cache their kills by hanging them on a long thorn—as a shrike, a slightly larger crow-size gray bird, will do—or in the nest cavity. They can catch their prey midair with an undulating flight pattern like a woodpecker's, a rapid series of wingbeats followed by periods of soaring with wings tucked up.

As a means of warding off predators or the small birds that are frequently seen mobbing Pygmy Owls near backyard bird feeders in winter months, these stealthy hunters have evolved two black spots with white rings around them on the back of their heads. This gives the appearance of two eyes continually watching behind them, much like the eyespots on the backside of a tiger's ears.

The world was layered in fog on this chilly morning as we wandered around and through the deciduous forest of Chinook Bend. We met a number of birders and photographers with the same long camera lenses and spotting scopes, looking for the same small owls. We met a man named Matt, covered in camo from head to toe, including his camera. And Gary, a tall older man, with glasses and a blue skullcap, who'd been out at least twenty times photographing these two owls. He'd spent so much time watching them that he could almost tell the difference between the nearly identical owls. He said he thought he would be able to find at least one within about fifteen minutes.

I wanted to believe him as we stood under a group of small trees and he showed me some of the photos saved on his phone. "He's not afraid. He'll hunt right at your feet," said Gary about one of the little owls he had come to know.

I was glad to know there were others looking. The more eyes on the trees, the better chance we had of spotting one of these tiny, well-camouflaged owls.

When Roger was last here, he had seen one of the owls by the parking lot, perched in a pine tree on the small tuft of needles at the top. So we scanned those same trees again as we wandered.

We were walking along a path beside a shallow pond built by beavers, scanning the trees, when a very loud honking broke the muffled stillness. We stopped still in our tracks. It was coming from somewhere above us in the fog.

A Trumpeter Swan emerged from the fog, with its long, graceful neck and wide wings as white as snow. We listened to its honking as it passed over our heads and flew off into the distance. A second one followed a few minutes later.

"That was worth the trip," Roger said, smiling.

It was the first Trumpeter Swan I had ever seen flying, something I had looked forward to since reading *The Trumpet of the Swan* by E. B. White as a child and again to my own daughter. I was thrilled.

All in all, it had been a pleasant morning. It was fun to be in the field with other birders and photographers, and I enjoyed walking out in nature in this forest where I had never been. I decided that I wouldn't be too disappointed if I didn't get to see the Pygmy Owls. Then I saw a tree with something on it. Could it be a Pygmy Owl?

I had never seen a Pygmy Owl, so I didn't have shape recognition to go by, putting me at a disadvantage with this group. All I had to go on was the advice Gary had given, "Look for a bird the size and shape of a small fat robin." The thing I'd spotted in the tree turned out to be a large leaf. I reminded myself of the many times I had mistaken ocean waves for whales, as I kept looking. The sun rose higher in the sky, and the day grew older.

Just as I was beginning to give up, I heard, "Bird here!" It was Gary who found it. Everyone in the park who was looking for the owls ran to his position. He stood almost exactly where I'd met him, under the trees where he'd shown me his photos.

About twelve of us stood in a semicircle about sixty feet from the owl's tree. "If we were surrounding the tree, he would get nervous," said Gary, the ad hoc leader of the group from sheer experience with this owl.

I searched the tree. When I finally found him, I was surprised by his diminutive size. He looked nothing like a fat robin. With his feathers puffed against the chill of this February morning, he looked much more like a fuzzy tennis ball with a tail, as wide as he was long. Yet there he was! My first Pygmy Owl.

Next, I looked for the owl through the spotting scope on the tripod that Roger had brought. I wanted to be sure to get a good look before the much-anticipated Pygmy Owl flew away. I had given up trying to take pictures. The camera equipment was too heavy for me to use easily, and I was spending too much time just trying to find the owl in the viewfinder. When I finally found the owl, he was against the tree with no silhouette, blending into the background. I was gaining a new respect for nature photographers.

The tiny Northern Pygmy Owl sat in the tree, feathers fluffed against the dampness. With the leaves still off the trees, the little owl had nowhere to hide, unless it was against the trunk of a fir tree in this thin forest. Little birds, like towhees, nuthatches, sparrows, and the occasional robin, twittered and cheeped nearby. The Pygmy Owl followed their every move, searching for its next meal. It glanced down into the brush below, staring intently, listening for the tiny rustling of an animal. It scratched its white-spotted head with the talon of its yellow foot. The owl looked hungry.

I carefully moved the spotting scope. Now I could see white dots of feathers against the brown ones on the tiny owl's head and body, yellow eyes, the feathers on its eyelids, and nostrils on either side of its beak. When it turned away from its onlookers, there were the Pygmy Owl's telltale black eyespots on the back of its head. If I had known how to tell if the owl was male or female, I could have done it in that moment. Through the lens of the scope I had an amazing view.

I also noticed the absence of the facial disk common to most owls. As daytime predators, Pygmy Owls don't need the sound-collecting satellite-dish-shaped face and asymmetrical ears of other owls. They rely more on their daytime sight than nighttime hearing. Yet I could see a slight dish-like indentation around the Pygmy Owl's eyes, as if it could funnel more light to its eyes to aid in vision the way other owls channel sound to their ears.

The little owl sat on the branch, surveying all around it with the ease of a lord surveying his lands. Pygmy Owls do have predators in larger owls and other raptors, as well as mammals such as weasels and cats, but on this day no predators appeared.

The owl stretched its left wing and with its sharp talons combed the wing, smoothing each out-of-place feather. I could hear the whir of motors opening and closing shutters on expensive cameras.

What is it about owls? I thought, as I began to watch the photographers almost as much as the Pygmy Owl.

The owl returned to a relaxed sitting position, with the fearless demeanor common to predators. He looked this way and that, calmly keeping track of any potential meals in the area. He also seemed to look at each one of us, without fear, right in the eye as if to say he would not withdraw from the challenge.

For many species, humans included, direct eye contact is a sign of strength, whether or not it is intended to be a challenge. But it is also a sign of connection. With their forward-facing eyes, owls appear to look at us directly. And it is no small thing to feel as if a wild animal has made a connection with you.

All too soon, our time was up and we took our leave of this little owl and its followers. It had been a good day with the added benefit of meeting my first Pygmy Owl.

Stephen Hiro, a heart surgeon from Tampa, Florida, fell in love with the Montana wilderness during a ski trip. "It was so exhilarating," he told me. "I decided I had to get out West."

The idea stuck in his mind until, without good reason, he found himself applying for a license to practice medicine in Montana. Then, several months later, he got a call from a heart surgeon friend with a practice in Missoula who knew about his newly acquired Montana medical license. His partner had cut off two fingers on his left hand with a table saw. If Steve could be there in two weeks, he could have a job. Steve said yes immediately and commuted for two years between Tampa and Missoula before finally making the move permanent.

As a heart surgeon, Steve worked long hours in a safe, dry, and temperature-controlled environment, but he'd always been interested in the outdoors. In 1995 at a charity auction, he bid on a "day in the field" trapping Long-eared Owls with Denver Holt of ORI. It was wet and muddy the day they went out, but Steve said, "I was fascinated by the whole thing. Denver and I just clicked, and he continued to allow me to participate on weekends with the research."

Now this retired heart surgeon turned citizen scientist has taken on researching Northern Pygmy Owls for ORI. I met Steve (on the same day I'd met Denver Holt and Matt Larson) one early spring afternoon while looking for Long-eared Owl nests near Charlo, Montana.

"Saw-whets and Flams are easy to find," Steve said. "You just bang on the tree and the females will look out of their cavities. But Pygmies are harder. You have to see the pair interacting to find the nest cavity. With Pygmies, it takes a long time. You never do anything in a hurry," he said, explaining the labor-intensive search.

For the last five years, he has begun his work in early to mid-March, first identifying possible nest sites. He starts with checking previously used nest cavities from prior years, then moves on to other good-looking sites, using the recorded call-and-response playback technique many owlers use to find owls. Once he gets a response, he begins phase two, hoping to see the male catch a vole and take it back to the female in the nest cavity during the pair-bonding phase of Pygmy Owl courtship.

It was during this pair-bonding phase that I went out into the field one early April morning with Steve and Matt Larson, who I'd searched for Snowy Owls in Barrow with several months earlier, to find Pygmy Owls in the forested outskirts of Missoula. I felt more intrigued than deterred by Steve's earlier warning about the difficult search. We met at five thirty, knowing that Pygmies were most active during dawn and dusk at that time of year. From the trailhead, we hiked about a half mile through the dark forest to the site Steve was watching, a keyhole-shaped nest cavity in a bare snag. I found a place to sit as we waited, watched, and listened for the Pygmy Owl's slow, rhythmic, whistle-like *toot . . . toot . . . toot*, similar in pitch to but slower in tempo than the Saw-whet Owl's persistent *tooh, tooh, tooh, tooh, tooh* and more whistle-like than the hollow woodwind-like *hoop . . . hoop . . . hoop* of the Flammulated Owl.

As the day grew brighter, we still had not seen any sign of a pair, and I remembered my earlier failed Pygmy Owl searches. I now believed Steve when he said that Pygmy Owls took some time and dedication to follow. Steve was determined to find a pair this morning, so we continued on to the other nest site he'd been watching.

Two years earlier Steve and Matt had managed to take photographs of Pygmy chicks as the down-covered babies grew first the follicles of feathers and then fully developed feathers. Then all of the chicks left the nest cavity together, fledging on the same day. Steve and Matt believed they had the first-ever recorded evidence of "synchronous fledging," something unique to the species of Northern Pygmy Owls.

Female owls usually lay their eggs asynchronously, one every one to two days, and begin to incubate those eggs as soon as they're laid. That means that in most owl families, there are chicks of a variety of ages. The older chicks are more fully developed and more capable of surviving than the younger chicks, which lag behind in development and struggle to compete with the older chicks for food. Some scientists have proposed that this asynchronous hatching may be due to the practice of "obligate siblicide," when the older chicks use the younger chicks for food in times of stress and starvation. This, it would seem, is one of nature's ways of guaranteeing the survival of at least some of the offspring.

Steve and Matt thought Pygmy Owls used a different survival strategy. They believed that Pygmies still laid eggs asynchronously but held off on beginning to incubate those eggs until all of the eggs were laid, resulting in synchronous development and synchronous fledging of Pygmy Owl young. But their pictures had captured only a single occurrence, much too small of a sample size to be anything other than a scientific note in a journal somewhere. What they needed now was more cases in their sample. Steve hoped they would find them this year.

To get the needed evidence, they would need to time their research precisely, arriving at each nest cavity on the exact day of fledging. To do that, they would need to know on what day eggs were laid. And if they wanted to band these Pygmy Owl chicks, they would have only this one day to catch them before they flew away out of reach. For now, though, the most important task on this day was to try to confirm that a Pygmy Owl pair was using this cavity to nest.

Steve wanted to look at the inside of the cavity to see if there were any eggs laid yet, so he carried a tall ladder, while Matt carried a borescope camera, used for examining drainpipes, which he would duct-tape to a long pole. We walked along in the early morning hours until Steve pointed to the tree with the second suspected nest cavity.

We listened closely for the calls of the tiny Pygmy pair. Matt and Steve heard the quiet call first. "Did you hear that?" Steve asked in a whisper. I shook my head no as Matt tried to find the owl. When he found it, he pointed in its direction. Steve then showed me where to look.

I spotted the tiny owl through my binoculars and again was amazed by the small size of this fierce predator. *Toot . . . toot . . . toot*, called the male to his larger mate, about the size of a ponderosa pinecone. Steve had found the female Pygmy sitting in the sun in a tall ponderosa pine, all puffed up on this chilly morning.

I could see the white spots decorating the male owl's head and wings, and the brown bars down his white chest. As his head swiveled in owl fashion, I could see the dark eyespots on the back of his head. Then the male flew to the tree with the nest cavity, clinging to the side of the tree like a nuthatch. From where I stood, he looked about the same size as a chickadee. The cavity's opening looked no larger than a fifty-cent piece, yet the male Pygmy Owl was small enough to fit inside. One reason the chicks had to be caught on the day they fledged was because the hole was too tiny to reach into with a hand.

The male ducked inside, turned around, and stuck his head out of the cavity, inviting the female in. The female, however, looked much too relaxed to move from her cozy position in the morning sun. The male spent only a moment more at the nest site before he flew to where the female sat. Then the pair copulated in the morning sun.

"Pair bonding," said Steve in his calm, cool doctor voice. "That happens frequently during this time. If she doesn't have eggs yet, she will probably lay soon."

Then the male flew off, possibly to find his love a tasty vole for her morning meal.

Steve and Matt decided that it was safe now to take a look inside the cavity. They leaned the ladder against the side of the tree and turned on their monitor to test the camera before sending it in.

Matt carried a pole with the camera taped to the end of it up the tall ladder. He was still ten feet shy of the nest, so he cautiously stepped to the top rung of the shaky ladder with one arm around the tree while the other held the pole and camera. According to Steve's observations, we knew that Pygmy Owls prefer to nest in cavities made by woodpeckers at least forty feet up in larches, also known as tamaracks; cottonwoods; and aspens. This nest looked to be at the Pygmies' desired height.

Carefully, Matt directed the camera into the cavity. On the monitor, it looked deep and dark. As Matt moved the camera around, a slight shelf appeared. "I don't see any eggs yet," said Steve, who was also looking at the monitor. "Probably a few more days."

Satisfied with the nest check, Matt climbed down and they packed up their gear. All except the ladder, which they hid behind a fallen log for another nest check in the near future. I took one more look at the female, still basking in the glow of the sun.

A few months later, I received an e-mail from Steve: "We were able again to serially video record seven chicks' plumage development and again it was completely synchronous. All 7 chicks fledged as well the same day and were able to fly immediately. Neat stuff."

Neat stuff indeed. Steve had his data showing more evidence that Pygmy Owls have a different type of nesting strategy than other owls—synchronous hatching and fledging of the young. Through long patient hours of observation, a citizen scientist was adding to our knowledge of the nature of Pygmy Owls.

The Long and Short of Eide Road

Long-eared and Short-eared Owls

The Skagit Valley, nestled between the forested slopes of the Cascade Mountains and Puget Sound, about an hour north of Seattle on I-5, is famous for its fertile soil, turned into bountiful farmland. Tulip fields awash in rich pink, yellow, red, and purple hues delight visitors in springtime. In summer, many people sample the bounty of produce at roadside farm stands on the way to weekend getaways in places like La Conner, Burlington, and the San Juan Islands.

But in the fall and winter, attention turns to winged visitors, migrants from as far away as Russia's Wrangel Island and the Alaskan and Canadian arctic tundra. Birders bedecked with binoculars, spotting scopes, and cameras with gigantic superzoom lenses wander the farm fields and tidal flats in search of Snow Geese, Trumpeter and Tundra Swans, Dunlins, Black-bellied Plovers, Marsh Wrens, Snow Buntings, Tree Swallows, Wood Ducks, Virginia Rails, Bald Eagles, Rough-legged and Red-tailed Hawks, Northern Harriers, and Long- and Short-eared Owls. They tell of life-list sightings in reverent tones at Skagit Valley birding meccas like "the West 90" and Eide Road.

I'd heard the rumors of Long- and Short-eared Owls over-wintering at Leque Island's Eide Road, between Stanwood and Camano Island. So one February day I decided to make the trip, hoping I'd chosen a time between one of the Pacific Northwest's famous winter rainstorms. To this point, I'd always gone with a scientist studying owls or an experienced birder intent on finding one. Now I would test my newfound owling skills on my own. A birder I'd met while watching the Snowy Owl on the Edmonds waterfront had told me that Short-eared Owls were easy to spot at Eide Road, that "they were right there on the sign by the parking lot." And while I knew this was possible, my wildlife experience told me not to count on an easy sighting.

"Turn left after you cross the bridge to Camano Island with all the construction," I remembered my wildlife photographer friend Roger saying. He'd been there earlier in the season to photograph the owls. As I neared the metallic gray wildlife cutouts of a bald eagle, a heron, and a salmon mounted as public art on the edge of the bridge, I looked for the Eide Road sign and made the turn through the construction zone. A short hill gave me a wide view of the Stillaguamish Delta into Port Susan Bay. A pair of white

Trumpeter Swans picked through the remains of an empty cornfield while a Red-tailed Hawk flew from its perch on an electrical wire running alongside the road.

There was only one other car in the parking lot. *Good. Peace and quiet*, I thought.

I wandered around for a few minutes, exploring the muddy trails up and over a dike overlooking the Stillaguamish River. Standing on a log and watching the river sweep past, I noticed the stillness of this rainy day. I stood quietly and breathed deep the earthy scent of rain and wood as I scanned the river's edge with binoculars. Seeing no owls roosting on any of the driftwood tree stumps, I slipped and slid my way back to the road as the raindrops quickened. Along the edge of the marsh, flocks of Dunlins, little long-legged shorebirds, probed the muddy puddles with their long bills for a tasty meal.

Farther along the road, a Northern Harrier, with its long barred tail and a visible white spot on its rump, crossed behind me to fly over the marsh. The Dunlins did not startle, so I guessed the harrier was after small rodents. The harrier was a good sign. Long- and Short-eared Owls are often seen hunting in the same vicinity.

A Bald Eagle sat in some variety of leafless deciduous tree, surveying the goings-on in the marsh below. While looking at the eagle, I missed the sign warning visitors of the hunting season currently under way. It had never occurred to me that two opposite activities—bird-watching and bird killing—would be allowed on the same piece of land. For the better part of the day I walked around in my black raincoat, sans orange hunting jacket, blending in with the natural backdrop, oblivious of the danger.

Just then, wild sounds rushed toward me. Wings beat the air as something large lifted off in flight out of the grass, breaking the stillness. My breath caught in my throat, and my heart pounded in my chest as I stepped back. The bird flew a few feet ahead of me, displaying its purple-and-green feathered head, white-ringed neck, and long brown tail. My breath released, and I relaxed. Only a startled pheasant, I realized, flushed from its camouflaged roost in the tall grass flying to safer ground. *Safer from what?* I wondered, looking around.

Fortunately for me on this day, there were no hunters around. After recovering from being startled by the pheasant, I noticed two women walking my way. Most likely the owners of the other car, I surmised. I saw them looking through binoculars and guessed they were also birding, so I planned to ask them about their sightings. As they passed a large fir tree, one woman touched the other's elbow, stopping her. It was a moment of discovery; she'd spotted something.

The two stepped back and set up their spotting scope before quickly relocating in the scope whatever it was they'd seen. I was curious, thinking there must be something exciting in the brambles surrounding the tree. I could see but not hear them, whispering with the thrill of an exceptional sighting. I walked slowly so as not to startle this mystery bird. One of the women noticed me and motioned for me to come. I continued my slow steady pace, even though I wanted to run.

"What do you see?" I whispered to the woman who had invited me, as I scanned the brambles with my eyes.

"A Long-eared Owl. Take a look," she said as her friend stepped back from the scope.

I looked into the viewfinder, and there sat an owl on a long bare branch, within a small grotto created by the tangle of blackberry and holly bushes. I looked up from the spotting scope to see where this medium-size owl was with my own eyes. But the owl was invisible.

"I see it in the scope, but where is it? I can't see it," I said.

"Just above that red bit of garbage," the woman said, directing me. I saw a discarded red popcorn-like box lying under the tree and looked just above it. *Humans*, I thought, shaking my head at the carelessness before returning my focus to the owl.

Then slowly, as if by magic, the owl appeared out of the background, like one of those 3-D stereogram posters where the pictures pop out at you the longer you stare at it. The bird was much lower to the ground than I'd expected.

"Wow! Great spotting," I said. "It blends in so well with the background."

Indeed, the chest feathers of the Long-eared Owl, mottled light gray with long and thin broken black bars, mingled with the dim, dappled light showing through the thicket in which she hid, concealing the owl from the casual observer or any predator that may happen upon it. The dish-shaped face was more oblong on the Long-eared Owl than on the Great Horned Owl, which the Long-eared closely resembles. Tan feathers and a dull-white *X* between the eyes added to the bird's highly camouflaged appearance. Subtle patterns on the dark-gray wings and head gave the impression of the bark of a tree.

Her long black ear tufts, from which the Long-eared Owl gets its name, were relaxed, falling open to about a sixty-degree angle. Had the owl been at all alarmed by our presence, her ear tufts would have stood erect atop her head. Instead, the owl opened one orange eye and watched our movements as she clutched her latest kill, a large rodent of some type, in one tawny-feather-covered talon. For the next several minutes we took turns looking at the owl through the spotting scope, which magnified the comb-like structure of each individual feather.

"Thank you for sharing that sighting. I never would have seen it on my own," I said. I probably would have been looking up into the trees, not yet understanding, until I could observe the owls myself, that this "edge species" prefers to nest and roost in thickets near open-field hunting grounds where it can glide and scan for food rather than using the other owl strategy of perch and pounce.

The Long-eared Owl (*Asio otus*) is found around the globe, from northern boreal forests of Canada to as far south as subtropical Mexico to Europe and Asia. The Long- and the Short-eared Owls (*Asio flammeus*), are the only owls from the genus *Asio* in North America, out of seven species and eleven subspecies worldwide. A subspecies of the Short-eared Owl, the white owl known as the Pueo (*Asio flammeus sandwichensis*), is found in Hawaii, where it's been for hundreds or possibly thousands of years.

On the wing, these two medium-size owls are difficult to distinguish. Both species tend to move from place to place, sharing a behavior known as "prey-induced nomadism," and both prefer to hunt in open grasslands, following the cyclical population of their favorite small mammal prey. In 2009, the Canadian Museum of Nature recognized a hybrid produced by the Long- and Short-eared Owls, a sign of their genetic similarity.

Because these two related species share similar habitat, they also share a similar taste in small mammal prey. One of the best ways to discover what owls eat is for biologists to examine the pellets that owls regurgitate every six to eight hours. Each pellet may contain several animals that the owl has caught, killed, and swallowed. Because owls swallow their prey whole and cannot digest things like fur, bones, and teeth, the skeletons are remarkably intact and can be easily reconstructed and identified—much like reassembling dinosaur bones, but on a much smaller scale. Owl pellets have become a hot commodity in educational circles as they are used by biology students to dissect and identify the small creatures these nighttime predators eat.

Using this technique with Long-eared Owl pellets over the years, ornithologists have reported on 813,033 prey items eaten by these owls worldwide. A review of all these scientific studies by ornithologist Simon Birrer revealed 477 different species listed as prey for Long-eared Owls, including 180 small mammal species, 191 small bird species, 15 reptiles, 83 insects, seven amphibians, and one fish. Small mammals accounted for 93.3 percent and small birds 6.4 percent of species taken. In North America, while 133 different species were found in pellets, members of the genus *Microtus*, voles, were most favored. Long-eared Owls wander open fields to look for a variety of vole species to satisfy their appetite, in part because each vole weighs, at most, one and three-quarters of an ounce. Short-eared Owls, while sharing the Long-eared preference for voles, are also known to take small birds such as terns and small gulls, especially when hunting near the coast and in marshy areas. Inland they may hunt birds as large as meadowlarks and blackbirds.

Long- and Short-eared Owls also share the habit of communal roosting, with Long-eareds taking the practice to the extreme. In northern Serbia, some twenty thousand Long-eared Owls are said to roost during the winter months. In the small town of Kikinda, Serbia, known as the "Owl Capital of the World," near the Romanian border, as many as eight hundred Long-eared Owls have been counted roosting in the trees of the town square. The owls have become such an important tourist attraction that the local government passed an ordinance banning any activities that would disturb the owls, with heavy fines for disobeying. The owls appear to be attracted to the large local population of meadow voles, which seem to thrive under the traditional farming practices used by local farmers.

"It's hard to work with the owls there," David Johnson once told me. "As soon as you get the net set, there are fifteen or sixteen owls caught in it." When working in Kikinda, you have to work fast.

In other parts of the world, like the northernmost United States, researchers are working fast for another reason. A thirty-year study done by ORI in Montana shows the Long-eared Owl is declining in their research area. Now they are working to solve the mystery of this decline before it's too late.

On Eide Road, both Long-eared Owls and Short-eared Owls caused quite a stir among local birders in the winter of 2014–2015, when the sightings were almost guaranteed. There was even a rumor that one Long-eared Owl pair that nested at the site the year before had returned. Even with the near guarantee of this season, I was still astonished by my sighting and grateful to have seen the Long-eared Owl.

"Have you seen any Short-eareds?" I asked the women on Eide Road as they packed their scope.

"We saw one flying. Not here but at another spot nearby. Then we never saw it again."

I stayed with the Long-eared Owl for a few more minutes before deciding to look around some more. Long-eareds are beautiful owls, and I'd loved seeing this one holding the catch of the day. Yet, truthfully, I felt a bit disappointed that once again someone else had pointed out the owl to me, and I was somewhat discouraged by the Short-eared Owl report.

I wondered if I would be able to find a Short-eared Owl on my own. Over the last year, I had learned a lot about owls and I was eager to put my knowledge to the test. But had I learned to think like an owl? If I could find one in its natural habitat, I would consider my owl quest a success. As with the Long-eared Owl, I still needed to learn about the true nature of Short-eared Owls.

I would learn more about the true nature of Short-eared Owls from Matt Larson, the ORI research biologist I'd worked with in Barrow, Alaska.

The ORI field station is on a small farm adjacent to the Ninepipe National Wildlife Refuge near Charlo, Montana, in the Flathead Valley just north of Missoula. I drove up the gravel driveway to the farmhouse, red barn, and other outbuildings now serving as scientific laboratories, all set against the backdrop of the striking snow-covered Mission Range.

Before Matt Larson and I headed out on this evening's Short-eared Owl survey, he showed me around the peaceful farm turned organized research center. The barn held research equipment, like various forms of owl-capturing nets and camping equipment. Another building contained a makeshift lab filled with jars, test tubes, and microscopes, along with cold storage and several species of stuffed owls sitting around. Still another offered a comfortable room to write papers or ponder all the unsolved mysteries of owls. The house, filled with books, maps, and research papers, provided office space.

We piled into his truck, a far cry from the four-wheel ATV we'd ridden in Barrow, which was much appreciated, as it was early spring and still cold in the evenings in Montana. We drove along gravel roads in between fenced fields of crops, cattle, and open space. Matt wanted to survey a couple of fields managed by the US Fish and Wildlife Service and some private property where ORI had permission to band owls, to see if any Short-eared Owls were using the land. Much of the land was used for grazing cattle or growing crops.

Like Snowy Owls, Short-eared Owls lay their eggs and raise their young broods of five to ten chicks on the ground in shallow bowls they've scratched out in the tall grass. Because of the danger inherent to ground nesting, the young begin to wander from the nest as soon as twelve days after hatching. But tractors needing to plant fields of crops or mow hay and herds of cattle tromping through the fields do not make for safe nesting habitat for Short-eareds.

In the United States over the last hundred years we have lost 99 percent of our native prairie habitat to farming and development. Of the remaining 1 percent, only 2 percent of that is managed to protect wildlife like Short-eared Owls. With all the development of prairie habitat, Short-eared Owls have seen a nearly 70 percent reduction in population across their known range in the United States. We are losing our Short-eared Owls.

Matt stopped along the side of the road, grabbed his binoculars and clipboard from the truck, and explained the survey protocol as we took our places beside an open field. "The goal is to establish some kind of baseline for this population," he said. Matt had developed this protocol after first trying audio playback, using the owl's own call to lure it in to be counted, as David

Johnson does with Burrowing Owls. The problem is, Short-eared Owls do not defend territories in the same way Burrowing Owls do, because of their nomadic nature. In his surveys, Matt was able to count 220 owls by simply scanning visually using binoculars and only 20 using the playback technique.

"We will survey the field for five minutes before moving on. We'll get as much done as we can. We only have a short window between when they start hunting and dark." He paused to take notes on the general environmental conditions, such as wind speed and cloud cover and precipitation or lack thereof, factors that could affect our ability to see owls or impact their propensity to fly. Owls prefer not to hunt in rain. They have given up the oil found in many feathers, which protects other birds from rain, for soft, silent feathers more valuable for stealthy hunting.

"Have you seen Short-eareds flying?" Matt asked.

"Once," I replied.

"They're pretty distinctive once you spot them. They sort of glide along the ground. Flap, flap, flap, glide, and hover," he said. "There's one there. Just over that small hill." He pointed in the direction of the owl.

I looked through my binoculars. "I see it!" I said, excited to see a Short-eared Owl in its nesting habitat.

"Good. You watch that one, and I'll look around for some others."

As I watched the owl hunting, I wondered about her life to this point. What places had she visited in her short owl life? Over what mountains had she flown in the spring rush to nest?

They are also specialist predators, meaning that they often rely on only one species of vole for their entire diet. In years and areas when their favorite voles are abundant, they congregate in great numbers and breed with abandon. Yet they rarely return to the same place.

One summer when Short-eared Owls were in greater abundance than they had been in ten years in western Alaska, several owls were radio-tagged. In winter most of the owls moved south across the western continental United States and central Mexico, while several spent the winter in the Great Plains. Of the three owls that kept their radio tags during the following year, none returned to Alaska.

In another study, Short-eared Owls tagged in Alberta, Canada, wintered in Montana and Kansas and returned the following year to within 125 miles of where they were tagged, somewhat loyal to the region but not a particular site. Because of their low site fidelity, the population of Short-eared Owls is difficult to determine. If not for the tireless efforts of owl biologists and volunteers, we could lose them and never even know they were gone.

When five minutes was up, Matt recorded the small number of Short-eared Owls we'd seen. It was difficult to say for sure if one of the sightings was one or two, because the bird continually disappeared and reappeared from behind a hill. The most accurate thing to do when there is a question about numbers during a survey is to record a range, so Matt recorded seeing two to three.

The result was not surprising. The then-vacant field had been grazed by cattle two years before. The ground needed more time to recover, to regrow the long smooth bromegrasses and bunchgrasses that Short-eared Owls prefer for nesting. Matt believes that a field in its fourth year of standing empty with no grazing would be the best for owls. But the government gets paid for selling grazing rights, and many farmers say they can't afford to leave their field fallow for four years. In this way, humans have a great impact on the loss of habitat for Short-eared Owls.

We surveyed for the next hour and a half until it was too dark to count any owls on the wing. We'd seen five to seven Short-eareds in total.

Several months later, I received an e-mail from Matt. "Tough year for the Short-eareds. We've found a handful of nests, but they've all failed. Not sure why." Sometimes even the scientists can't explain the workings of Mother Nature.

It may have been a loss that year for the Short-eared Owls in the Flathead Valley of Montana, but for Matt Larson the surveys were a success. Biologists from Idaho, Nevada, Utah, Minnesota, and other parts of Montana agreed to begin using Matt's technique the next year. With many people watching, maybe we can learn how best to help the population of Short-eared Owls to recover.

Back on Eide Road in the Greater Skagit and Stillaguamish Delta this February day, with my trip to the Flathead Valley still months ahead of me, I did not yet understand the true nature of Short-eared Owls. The rain began to fall harder, so I started to walk back to my car. On my way, I met two more women out for a day of birding. Now it was my turn to point out the Long-eared Owl that was still sitting in the small grotto, protected somewhat from the now-steady rain.

"Oh, you're so beautiful," one of the women cooed quietly to the owl as she studied its rich colors. "Thank you for pointing her out," the woman said to me. "I'm going to make a glass piece of you," she said, looking toward the owl again. "I blow glass."

"That sounds amazing," I said. We talked for a few minutes more about her work before she told me she often comes to Eide Road to walk and to find inspiration for her art. I told her of my quest to find a Short-eared Owl.

"Oh, they're here," she said. "Go down to the end of the dike and stand up there. Watch around three thirty or four o'clock, and you'll see plenty. I've seen as many as six or seven of them flying around."

So if I was going to see Short-eared Owls on this day, I would need to wait a couple of more hours until the owls began hunting. I watched the aerial acrobatics of a Northern Harrier in the field before me as I ate lunch in my car. With the variety of raptor species I'd seen today and the presence of both the Long-eared Owl and the harrier, I guessed hunting was good. I began to feel better about my chances of finding a Short-eared Owl.

The stillness of the day once again surrounded me. With the windows down, I enjoyed the sound of the rain falling, gently tapping on the roof of the car and on the ground outside. It was ten minutes to three, and by my calculations, I had another hour or so until the Short-eared Owls began to hunt. I scanned the field in front of me for other bird species wintering at Eide Road.

Then suddenly, I caught some movement out of the corner of my eye. If it was a bird, I couldn't tell. There was no wild fluttering of wings as there had been with the frightened pheasant. I glanced in that direction to see an owl, a Short-eared Owl, sitting on top of the sign right beside my car. I could scarcely believe my luck. It must have flown up from the ground, where it had been roosting all along, nestled down in the long brown grass, completely camouflaged. All my preparations and study of Short-eared Owl habits and where to find them had mattered little. In the end, it was the owl that found me.

It sat on the sign in true owl fashion, its beige head with dark-brown stripes and coal circles around its eyes reminiscent of an Egyptian pharaoh. It swiveled its head 270 degrees, scanning the ground for a meal. The short ear tufts atop the owl's head, for which the Short-eared Owl is named, remained invisible.

The hunt was on, and the owl did not stay long on the sign. I leapt from my car, binoculars in hand, to follow the owl's flight. Short-eared Owls are not perch-and-pouncers, like Saw-whet, Spotted, and Barred Owls, but prefer to hunt on the wing. The Short-eared soared out over the field, occasionally slowing its speed to the point that I thought it would fall out of the sky. Yet it continued to soar silently, using its satellite-dish-like face to funnel the tiny sounds of scurrying voles and field mice to its ears. When it thought it had pinpointed the location of some prey, the Short-eared dropped face-first into the grass, extending its long sharp talons at the last minute to make a grab. The owl missed again and again before it moved to the other end of the field. I followed as close as I could, watching through binoculars.

Over the sixty-seven million years that owls have been on the earth, they have evolved several unique features in their feathers, different from other bird wings, which enable the silent flight of these efficient predators. Much like the wing of an airplane, the wings of most birds create turbulence and noise as the air flows around the wings, allowing birds like the pheasant to create quite the commotion on takeoff, perhaps to frighten predators. Owls have developed a specialized comb-like structure on the leading edge of their tenth primary flight feather—or the first finger if you imagine a wing as a hand—that reduces the air pressure along the wing. Special fringe along the trailing edge of the owl's flight feathers breaks the turbulence into many microturbulences, thus quieting the feathers in flight. In addition, on the dorsal, or top, of each soft owl feather, a fuzzy porous structure that looks like a tiny forest under a microscope, called the "pennula," further assists in the owl's silent flight. Scientists are hard at work studying these specialized, evolutionary steps made in the owl's quest for silent flight as a model for both airplanes and wind-power generation.

I hiked to the end of the trail, following several Short-eared Owls now hunting over the open fields of Eide Road. In all, I counted at least six Short-eared Owls. I knew it must be a special place to sustain so many large avian predators.

The sun was low and dropping fast as I turned to head back to the car. As I neared the grotto of the Long-eared Owl, I stopped and turned back for one last look at the Short-eareds hunting on the wing. As I turned, I noticed two orange eyes staring right at me from a tree branch just higher than my head—not more than ten feet away. I stared back, unmoving, for an unexpected moment of connection with the wild Long-eared Owl.

In her wild eyes, I did not see a "bird" anymore. I realized in this moment that I'd made some distinction between owl and bird, giving the owl more value and meaning somehow. For me, the owl had become another way of knowing something more about my world than I had known before. It had become a symbol beckoning me forward toward a larger view of my place as a human in this natural world. I had been on a journey with the owl over the last year. Was I any closer to discovering the owl's message? Was I any closer to thinking like an owl?

I enjoyed a few last moments of quiet as I walked back to my car. Several more Short-eared Owls flew over the field. The harrier I'd seen earlier also hunted in these last moments of daylight. A long-legged Dunlin walked along digging for food. A Bald Eagle sat watching over all. This open-field ecosystem with all its players—predator and prey moving together and apart—seemed intact. At this moment, all was well in the owl world.

Thinking Like an Owl

Great Gray Owls with Great Horned Owls

High in a Douglas fir, in a natural bowl created by the upward rather than outward reach of the tree's branches, an abandoned Cooper's Hawk nest holds a large gray owl. With her large gray-feathered head and classic round, flat owl face, reminiscent of a bull's-eye with its gray-and-white concentric-circle pattern, she seems otherworldly in her stillness—*in* the world but not *of* the world.

She listens and watches the forest moving around her through wise yellow eyes. Her dappled gray body blends perfectly with her forest home, and when unmoving, she is difficult to distinguish. Long tail feathers stretch gracefully over one edge of the nest, which seems too small to hold her. Underneath her, tiny lives grow and change in their protective shells for four weeks, warmed by their mother's body. She will not leave the nest for more than a few minutes at a time for the next month.

A raven's cry in the distance commands her gaze before the low call of her mate bringing a meal reaches her ears. She returns his call with a low *whoo, whoo* of her own, followed by an excited high-pitched cry. She's hungry. The male, sole provider for his family while his mate tends the eggs, glides silently to the branch just above her. As the pair exchanges squeaks and chirps, he carefully walks down the branch with wings outstretched, as if walking a tightrope, until the female can reach the rodent dinner he brings. She swallows the meal whole, headfirst. With more squeaks, the male flies off to catch the next course. To keep them both well fed, he must find five small mammals for each of them every day.

On the hunt, the male flies from the nest in the forest to a nearby meadow lined with young tamaracks. The native bunchgrasses of the meadow—basin wildrye, bluebunch wheatgrass, and tufted hairgrass—offer opportune hunting because of their wider plant spacing compared to the grasses of city lawns. The big owl with the large gray-feathered head that evolved to absorb the shock of hitting deep snow when hunting for prey twenty-four inches down, balances first on top of a small tree, then alights on a thin metal fence post. With intense concentration and patience, he listens.

His bowl-shaped face gathers the tiny noises from his prey moving about in the dried grass below. With precise triangulation of his prey's location, calculated by using his asymmetrical ears, he leaps from his perch. In silence he glides, then floats on a five-foot wingspan, pausing just above his target before he dives face-first toward the tiny vole, dagger-like talons outstretched before he strikes the vole with a force equal to that of a man being hit by a Mack truck. He takes his prey in his yellow beak and flies off in the direction of his mate. He performs this act over and over until they are both fed.

Once the chicks hatch, he will hunt day and night, as long as it takes to provide for his growing family.

In the northeastern corner of Oregon, stretching into southeast Washington, stands a mountain range called the Blue Mountains. Among the tall orange-barked ponderosa pines, the thin green-needled tamaracks—which turn a spectacular yellow in the fall—and the quaking aspens, Great Gray Owls nest. "If you want to find Great Grays, I would look there," said Washington bird expert George Gerdts when I'd talked with him about the owls I could expect to find in the Pacific Northwest.

I *did* want to find Great Grays. I'd spent a little more than a year learning all I could about the owls of the Pacific Northwest. Now I wanted to test my newfound skills to see how well I had learned to "speak owl." I didn't know it at the time, but Great Gray Owls are notoriously elusive, blending as they do with the forest that surrounds them. Finding them is almost always difficult.

I discovered the Spring Creek Great Gray Owl Management Area in the Wallowa-Whitman National Forest through Evelyn Bull, the preeminent Great Gray Owl biologist in the United States. She wrote the species account for the Great Gray Owl in *The Birds of North America*; for biologists and birders, that's akin to writing one of the books of the Bible.

Bull's interest in Great Grays began in 1982, when a forest fire ripped through the national forest near the Thomason Meadow Guard Station in Wallowa County, exposing for the first time a Great Gray Owl nest there. Evelyn set out with a fellow Forest Service biologist, Mark Henjum, to discover just how many Great Grays they had in the Blue Mountains. What they discovered was the highest concentration of nesting Great Gray Owls in the world.

Along with another small population of Great Gray Owls in California's Yosemite National Park, these owls in northeast Oregon are the last of the holdovers from a population of Great Grays that was separated during the last ice age from the larger population in the boreal forests of northern Canada. Along with a few other small pockets in Eastern Washington, Western Montana, Idaho, and southwestern Oregon, these small groups are the most southerly Great Gray Owls in the world.

From 1982 to 1988, Evelyn Bull and Mark Henjum studied 71 nesting attempts from 24 pairs of Great Gray Owls and their 107 offspring. They studied their preferred types of prey species, their movements—they seem to move within an area throughout the year, rather than follow the "to and from" migration patterns of some owls—their nesting behavior, and their habitat use, including their preferred nest height. They found that the Great Grays prefer to nest about fifty feet off the ground in Douglas fir, tamarack, or ponderosa pine in the abandoned nests of Northern Goshawks, ravens, or Red-tailed Hawks. With the loss of habitat for these nest-building birds, Great Gray Owls have been hard-pressed to find suitable nesting sites. So Evelyn began installing nest platforms for the owls to use, a program that has spread to other parts of Oregon.

"How can I find them?" I asked Evelyn.

"Call the Forest Service office," she said. "They should be able to tell you. They check the platforms as a courtesy to the members of the public who want to see Great Grays."

The owls are popular with people from as far away as Europe who travel to the Blue Mountains with the hope of seeing this magnificent species. In 2015, the Great Gray Owls of Spring Creek had one of their most successful nesting seasons in years, with six successful nests. But I did not see the Great Gray Owls of Spring Creek.

On Evelyn's advice, I contacted Andy Huber, founder of the Grande Ronde Overlook Wildflower Institute Serving Ecological Restoration (GROWISER), a nonprofit native plant preserve in the northeast corner of Oregon. Three weeks later, I arrived at GROWISER.

Since 1992, Andy, like a modern-day Aldo Leopold, has been restoring the 220-acre site to its natural old-growth state. As an agronomist, or seed scientist, and retired professor from

Oregon State University, Andy has focused on restoring the trees, grasses, and flowers native to the area. He's particularly interested in mountain lady's slipper, a native wild orchid species that only he has learned to grow from seed. He believes the native orchid is an indicator species reflecting the health of the forest, much like the Spotted Owl.

For the past three years, Andy had seen the Great Gray Owls occasionally on the property, but he had not put it together that they were nesting there. "I wouldn't have noticed the nest this year if I had not been working on the trail that goes by the lady's slippers near there," he told me. Andy leads tours of GROWISER for school groups, college classes, garden clubs, orchid societies, and now birders and nature photographers. The first time he was certain that Great Grays were on his land was on June 1, 2012. He was giving a tour to the Oregon Orchid Society, which was being filmed for Oregon Public Broadcasting's *Oregon Field Guide*, when they came across one of the owls. The photographer was so impressed by how human-friendly the owl was that he filmed her for ten minutes.

On my first visit, Andy took me into the forest near the field station to show me the Great Gray Owl nest. He lead me down a lightly worn path, past a very large smoky quartz crystal he had placed there to "ground" the "energy" of this place, and an "E.T. Visitor Center" sign with a small extraterrestrial pointing in the direction from which we'd come. I smiled at Andy's sense of humor but had to wonder if he was really joking.

We walked partway up a hill before turning onto a path into the forest. Fifty feet in, we came to a worn place at the edge of a snarl of Nootka roses. Grand fir stumps, cut to a height of six to seven feet, stood mixed with tall Douglas firs and naked tamaracks, which had yet to grow their needles for the spring.

"I want to re-create the old-growth forest that was here," Andy explained. "After they logged this forest seventy years ago, all these trees grew up to the same height. The forest was too thick, so I cut off the grand firs that spread most easily, to be sure the others would get enough water and nutrients. Also, they make great habitat for the Pileated Woodpeckers that will eat the ants that use the stumps." Later, Andy would marvel at the synchronicity that had led him to create these owl perches.

Then he pointed to the nest where a female Great Gray had been sitting on eggs since March 4, about a month before my visit. They were due to hatch any day. At first, the nest was invisible to me amid the jumble of forest canopy. As I peered with binoculars through the branches, Andy gestured toward the area where the nest should have been. "Stand right here," he said, pointing out the invisible line of sight from that spot to the nest. But the nest remained well camouflaged in a tall Douglas fir.

Finally, I discovered the tangled bunch of twigs and branches. Then, as if she had materialized out of the trees surrounding her, I met the yellow-eyed gaze of the Great Gray mother owl. I could see the owl's shape, large gray-feathered head, and body atop a nest that seemed much too small to hold her and the lives growing underneath her. She looked at me only momentarily, barely aroused from her life-giving meditation.

"I see her," I said as I wondered at the maternal dedication that would allow an animal to sit almost entirely still for weeks. What type of Zen state of being-ness must she experience?

"Do you know where the father is?" I asked Andy.

"I haven't seen him today," he replied.

Great, I thought. This was my chance to find a Great Gray on my own.

"He's got to be here somewhere. I think they roost pretty close to the nest," I said as much to myself as to Andy. I remembered how close the male had been to the mother and chicks on the afternoons I watched the Spotted Owls on the slopes of the Cascades and on the Olympic Peninsula. The Great Gray Owl (*Strix nebulosa*) shares the genus *Strix* with both the Northern Spotted Owl (*Strix occidentalis*) and the Barred Owl (*Strix varia*). So I was guessing that their forest nesting behavior would be similar.

As we continued along the path down a short hill, I scanned the trees below me. Seeing nothing but more trees, I moved slowly, looking up into the high branches, focusing my search on scanning individual trees. Although the task at hand, to find a very large owl sitting on a branch, would seem an easy one, Great Gray Owls are well hidden among evergreen boughs. I paused my search to look down at the ground to make sure I didn't trip. When I looked up again, suddenly there he was, a large gray owl—right in front of me—hardly fifty feet away, across the forest clearing, on a branch that was level with my eyes. He sat in peace, eyes closed. I am sure he heard us coming, as we spoke in normal voices during our search, yet surprise did not register on his face. But it did on mine. A wave of excitement rippled through my body. Triumph registered in my mind. In a simple way, I had stepped into the owl's world. I had learned to think like an owl.

Unlike my time on Eide Road, where I had been prepared to go in the opposite direction when the Short-eared Owl flew up out of the grass and found me, this time I found the owl. It may seem a small thing, but it was the small things, the small triumphs in small moments, that led me along this owl journey.

Pine needles cushioned the spot on the forest floor where I sat watching the male Great Gray, as if he were the Dalai Lama and I his follower sitting before him and awaiting his next lesson. The owl began to preen, pulling each soft gray feather gently through his beak, reconnecting the tiny hooks that allow the feather to create a barrier of warm air against the cold or support the owl in silent flight. Then he rested, eyes closed for a few moments, until he perked up at the next seen or heard curiosity, which my human eyes and ears could not or did not detect. Mystery sufficiently solved, he closed his eyes and rested again until a low *whoo, whoo* from the female in the nest behind me rousted him from his reverie. He stretched his wings, curling his back like a cat, shook first his body and then his head, slowly to one side and then the other, fluffing his feathers. Then he leaned forward on the branch and fell into flight toward his preferred hunting field, as his mate had apparently requested.

Andy and I walked down to the field, about one hundred yards as the owl flies. "He hunts down here in the evening," said Andy, who this year had begun observing the owls from the start of the nesting season.

The male flew between the thin metal fence posts Andy had placed in the field to hold the tin cans he'd tied on them to scare the deer away from the aspen seedlings he'd planted over the past couple of years. "I didn't know those would come in so handy when I put them in," he said with a smile.

It was hard to believe the big owl could balance on the top of such a tiny post, but he looked comfortable as he watched and listened for his next meal. Again and again he soared, floating on his five-foot span of wings, then hovered and dropped into the grass with such force that I could hear him hit the ground from across the field.

I was surprised that the male owl ever missed. Yet he caught a rodent in only one of every four or five tries. I'd seen him deliver meals to the mother in the nest, but this time he swallowed the prey himself to keep up his strength for his family. In a single year, one Great Gray Owl may eat as many as eighteen hundred rodents. That night we watched this male fly across the road to another farm, land on the posts, and hunt in the abandoned barnyard. We followed his movements through binoculars until we could no longer see him, losing him to the dark and the nearby forest.

Fascinated by the prospect of seeing baby owls, I went back to GROWISER ten days later as the chicks began to hatch. I found a comfortable spot to sit and observe the behavior of the Great Gray mother owl on her nest.

Given the time, I will always choose to watch animals in their natural habitat. To really learn about any animal I am interested in getting to know, it is important to me to observe their wild ways. I consider it an honor to witness behavior in the wild with the animals undisturbed by my presence. Each encounter becomes a fond memory, as if I'd met a new friend.

On this day, I met the mother Great Gray. It was still early spring in the Blue Mountains, and the weather was unstable. I wore my hat and gloves and a winter jacket against the cold mountain air. I zipped my layers and pulled on my hood as a cloudburst of hard snow pellets pelted the area. The mother owl didn't move in the spring storm. I marveled at her ability to keep herself and her tiny chicks—still incapable at this age of generating their own body heat—warm and cozy under her bare brood patch.

Then Andy arrived with a GROWISER board member, Shawn Steinmetz, an anthropologist for the Umatilla tribe, who had never seen the Great Grays. In the friendly conversation about owls that followed, I forgot my chill.

As the three of us stood among the tall grand fir stumps watching the mother owl in the nest, Shawn told of an encounter he'd had with a Great Horned Owl chick a few years back. He had rescued the chick when it fell out of its nest. It couldn't climb back up the tree on its own, so Shawn built a platform for it about ten feet up in a tree and placed it there. The mother continued to feed the chick, and the baby survived.

I remembered the Great Horned Owl mother and chick I'd photographed on the Umatilla Chemical Depot the spring before when I'd worked with David Johnson on Burrowing Owls. My friend Claire Tahon from Belgium had found the chick sitting in her "nest" in the notch of a large oak tree where the trunk split into three branches. There was no nest material, like feathers, to line the nest, and no branches to shape it, yet the baby seemed secure and comfortable as she peeked at me over the edge of the shallow bowl.

Great Horned Owls are one of the most widespread owl species, occurring in nearly every habitat in North America. Subspecies range in color from rust to gray. Great Horneds typically nest as early as late December into February, depending on their choice of habitat, so it was a bit late in the season for a Great Horned Owl chick on the Depot, I thought, and the chick was nowhere near fledging.

Her mother, in a nearby tree, caught my attention with a loud call. She then flew from her perch tree to another tree several yards away, as if she wanted to lead me away from her baby. Her strategy worked like a charm. With my camera at the ready, I followed the mother, hoping to capture a nice shot. Through my lens, I could see the mother's mottled brown wings and head, the thin brown horizontal bars up and down her front, and the tall black feathers atop her head that give the Great Horned Owl its name. These "ear tufts," as they are called, are neither ears nor horns, but feathers, which aid in camouflage as well as revealing the Great Horned Owl's emotional state in the same way as the ear tufts of Long-eared Owls. As I focused my lens on the mother Great Horned's face, I noticed that she only had one eye. Her left eye was missing from the socket.

Now I understood why the chick was so young. Perhaps the mother had tried to nest earlier and failed. Or perhaps it was harder for this female to find a mate. With only one eye, her odds for survival were slimmer than usual. She and her mate would have to work harder to keep predators away and keep their chick alive. For now, the pair seemed to be doing quite well. If that kept up, another Great Horned Owl, the top nocturnal avian predator, would soon fly off into the wide world.

Being a mother myself, I wondered at the patience, determination, and skill it took for these owl mothers to raise their families in the owl way. What must it be like to watch your babies fly away forever?

Under the Great Gray nest at GROWISER, as we stood in the pelting snow, Andy joined me in trying to count fluffy white baby owl heads in a pile under the mother through the thick branches that hid the nest. It wasn't easy, but when the father brought a meal, the mother carefully tore off pieces of the rodent to feed to each opened mouth. Finally, through our binoculars, we were able to count, with 90 percent certainty, three chicks.

Once the chicks were fed, the mother rearranged herself on the nest and turned around to face away from us. How she fit into, or maybe onto, the increasingly crowded nest—with three squirming chicks snuggled underneath her—is a magical owl act beyond my comprehension. Then she began to dig in the nest with her beak. She poked and prodded and pulled at something beneath her. Was she trying to enlarge the small nest? Was she eating something? Was she helping another chick to hatch?

What I suspected but did not know at the time was that Great Gray mothers are very clean and careful with their nests. When I looked under the tree for owl pellets, I found only one, from an entire month of sitting on the nest. There was also no whitewash, the scientific term for owl poop, down the side of the tree as we would find under roosting sites.

It was a persistent mystery that Andy and I discussed on several occasions. Where were her pellets? We knew she was eating. We'd watched her swallow food. Was something eating them off the ground? Was the mother careful to expel them into the nest, so that no predators would notice it?

After careful observation, Andy would discover her secret. The mother owl flew each night to one of the nearby tree trunks Andy had sawed off. There she dropped a pellet and any other waste.

I also learned that mother Great Gray Owls eat the shells of the cracked eggs as their chicks are hatching, a bit like when a mother deer cleans the placenta from her newborn fawn. The mother Great Gray removes the pellets and feces from the nest until the chicks are able to fly.

After I'd left GROWISER, Andy kept me posted on the owls' progress with daily e-mail updates. One day, he sent this: "I am quite sure there are four chicks, instead of three. One is so small, though, that it is only about ⅓ the size of the other three." What Andy and I had witnessed that evening may have been the hatching of the fourth baby Great Gray Owl.

Baby Great Grays grow quite fast, growing in two to three weeks from downy white eyes-shut-tight, open-begging-mouthed bobbing heads to large fuzzy gray upright-and-steady, wing-flapping owl babies with dark-gray faces and fuzzy gray feet. They are restless and eager to explore their world. By the end of April, only about three weeks after my last visit, the Great Gray chicks at GROWISER had become branchers and began climbing out of the nest onto the branches surrounding the nest. They were always hungry and had become quite careless in their quest to be first to reach their dad and the food.

"Now there are three left in the nest. And one on the ground below the tree always begging for food," Andy wrote. "Seems a little early, but with all the wing flapping yesterday, I'm not surprised . . . I hope she/he survives the night. Lows are in the 30s with possible snow showers. Just thought you'd like to know."

I knew a baby owl on the ground was not a good thing, but I had heard of baby owls falling out of the nest and climbing back up the tree, using sharp talons and beaks to pull themselves up, so I let it go, knowing the baby's parents were there watching over it. Each day Andy sent a status report about the baby owl on the ground, and each day I expected that it would have climbed a tree. But on the fifth day, the baby owl still had not found its way back up. Now I was getting worried.

That night I woke up in a panic, thinking about the baby owl on the ground. A persistent image of Shawn, the GROWISER board member who had saved the baby Great Horned Owl, kept appearing in my mind. I could see him standing in the forest, talking with Andy about his rescue effort, as clearly as the day it had happened. Was I thinking now from the mother owl's perspective? I knew that the baby was too young to climb the tree. I knew it needed to be rescued. But how? I got up and wrote, concerning the baby owl's plight, to David Johnson, the expert on owls.

"A baby owl on the ground is never safe," David wrote back. "I would strongly urge your friend to pick it up and place it on a branch at least 10 feet off the ground. Make sure the adults are watching this happen."

I forwarded the message to Andy right away and waited impatiently the rest of the day, wondering if he had rescued the baby. Then came this evening update:

> Success! I looked for the chick on the ground and found it near the trail close to where we had seen it yesterday. It was alive but didn't seem to have too much energy. So I went back and got a long ladder. The mother watched from a nearby tree as I gently picked up the baby and carried it up the ladder. It was clapping its beak but didn't seem too disturbed. It calmed down nicely when I put it next to my body. It grabbed the branch easily and seemed very happy. Then I wanted to see if the mother would feed it. At about 7:15 the dad brought in another vole. The mother was extremely vocal. She used a much different language than I've ever heard her use. It was like she was saying, "The baby is on a branch, and I've got to go feed it!"

Many owls, and other birds for that matter, feed the largest chicks first because they are the ones clamoring the most for the food. They can push the others aside or intercept meals intended for the smaller chicks.

There is some evidence, however, to support the notion that Great Gray mothers take care of the weakest chick first. And in this case, Angel, as Andy named the chick (a "fallen angel"), which we would come to believe was male, was now the weakest chick. Over the coming days, as Angel was nursed back to health by his doting mother and father, many photographers visited the family and posted the photos online. The story of these Great Gray Owls spread around the Northwest, drawing over one hundred visitors from places like Portland, Corvallis, and Seattle.

As the other chicks began to climb out of the nest, the mother left the nest, but she roosted nearby so as to protect the still-growing chicks while the father hunted night and day. Angel fell a second time, and Andy rescued the baby again. From my home near Seattle, I read every update Andy sent, both excited and envious of his interactions with the owls in this developing story.

Wanting to see the chicks again before they grew up and flew away, I scheduled another visit for the middle of May. Five days before I was to return, I received a disturbing message from Andy: "The father was taken by a Great Horned Owl."

My heart sank. I reread the message. Did that mean the father was dead? What would that mean for the young owl family? Tears began to well in my eyes. Without the father, how would the mother raise all the chicks on her own? Could she catch enough to feed herself and the chicks?

I remembered a story that one of the biologists from ORI told me about a Great Gray Owl nest in Montana. The male Great Gray, with a vole in his talons, was killed by a car. Without the father to hunt for the mother and chicks, the nest failed. Would this nest of owls, my friends, fail? Would the chicks die one by one?

My next thoughts turned to the Great Horned Owl that had killed the father. Anger welled up in my mind. My first thoughts turned to violence to protect the Great Grays. *Kill the Great*

Horned Owl before he kills the mother and her chicks. I felt this thought work its way through my mind before I could think more clearly and my thoughts spun themselves to their inevitable conclusion. What would be the point? The Great Horned Owl is not to blame.

On my first visit, Andy had asked me if I knew if or how Great Horned Owls and Great Gray Owls interacted. I didn't know at the time that Great Horned Owls are predators of Great Grays. Even though the Great Grays are the tallest of the North American Owls, they are mostly feathers, weighing only about two and a half pounds. Great Horned Owls, third in size after Great Gray Owls and Snowy Owls, weigh three to four pounds and are one of the more aggressive species. To kill a Great Gray Owl would be a risky hunt for a Great Horned Owl, chancing the loss of an eye or a talon that would endanger his own survival, but sometimes they take the risk.

I remembered the one-eyed Great Horned Owl mother on the Depot trying to protect her chick. If the Great Horned Owls are desperate enough, they will attack the larger Great Gray. The Great Horned Owl father may have been that desperate because helicopter logging had been taking place near his territory, just across the valley from the Great Gray territory on GROWISER. Most likely the logging had disturbed the Great Horned father and his usual prey, so that he had to fly farther to provide for his growing family. Now, because his family and livelihood were threatened by human activity, he became a threat to the Great

Gray Owl family. Understanding this, was I willing to choose one over the other to make this "Sophie's choice" between species?

Andy immediately began e-mailing a local wildlife shelter and other raptor experts for suggestions about how to help the owl family. One idea was to take one of the chicks into captivity to raise as an educational bird, thus giving the mom a better chance of raising the other three. Another suggestion was setting up a feeding station. I remembered the mice I had seen offered to Barred and Spotted Owls. When those owls had chicks, they fed the mice to them. I thought surely the Great Gray mother could do the same. Again, I wrote for advice to David Johnson, who wrote back with his recommendations.

> It is the male who feeds them as they shift from branchers to fledging (and until dispersal). In this situation, the female may kick into gear and feed them to the extent she can. But supplemental feeding will help her. The young can eat about 3–4 mice per day (per owl) . . . so try to provide at least half of that supplementarily. Or as best you can. It will be most effective if the young have a central spot (nest) to use as a feeding depot. If the young are fledged now . . . then try to find them and put a dead mouse on a stick and feed

> them (or put the mouse beside them). If the young are jumpers, or within reach, you should be able to feed them by hand.

I forwarded David's note to Andy, rearranged my schedule, and headed to GROWISER to help feed the baby owls.

When I arrived, Andy took me to the site where the Great Gray father had died. Andy had been leading a small group tour of the property, when he'd seen the Great Horned Owl fly up from the grass, carrying something very large. As the Great Horned flew off, one wing of the male Great Gray Owl hung limply at his side. "I didn't know what I was seeing at first," Andy told me with the same visibly shaken look I'd seen from Jamie Acker when he told me about his Barred Owl friend Gus. Andy showed me a pile of gray feathers. "I just can't believe he's gone," he said.

In the owl forest again, I stared at the now-empty nest tree. The owl family was gone. The ten-year-old borrowed nest, where the owls had raised at least three years of owl offspring and where Cooper's Hawks had raised four families before that, had fallen.

Then I heard the chicks' food-begging calls, high-pitched, like distant orca cries in the sea, coming from the trees nearby. Where was Mom? Was she feeding her chicks? Was she feeding herself? Or did she continue to wait for her mate, who would never come?

Andy finally found her sitting alone on a low branch, unmoving, her eyes not bright yellow as usual but sunken in, grief registering all over her owl face. "She hasn't said anything in days," Andy said. "She used to talk constantly to the chicks. Now she just looks sad."

I could sense it from her too, the sadness over the loss of her partner.

"I'm sorry for your loss," I said quietly to the mother owl. I hoped she could tell that we meant to help.

Over this year I had been studying owls, I had come closer to the life-and-death struggles of these wild animals than any others I had observed: the Burrowing Owl's choice to leave her eggs to save her own life, the struggle for survival over the extinction of a species of the Spotted Owls, and now the death of the Great Gray Owl father, leaving his family to fend for themselves. What did it all mean? I felt there was something deeper here that I was learning or some message from the owls I was to hear.

As we walked out of the forest, my mind concocted different strategies for getting these owls fed. "I know she will take live mice, if we can find some to give her," I said. "I've seen other owls take live mice."

"I set a trap in the barn this morning," Andy said. "We can check it in a bit." It was the first of many live traps Andy would set as the Great Gray Owl saga progressed.

But we also needed to figure out how to get the mice to the owls. The chicks were far up in the trees by now, so David's idea of hand-feeding them wasn't an option. The nest had fallen apart, so we couldn't use it as a feeding station even if we could have figured out a way to get the mice up so high. So the only option left was feeding Mom and hoping that she would feed the chicks.

Or tying a dead mouse to a string on the end of a stick and feeding the chicks that way—reverse fishing, in a sense. With four chicks and herself to feed, she would need twenty mice a day at a minimum, we calculated. David had suggested supplementing half that number, so Andy would need to catch and feed at least ten small mammals a day.

Andy had one mouse he'd caught earlier in a mousetrap in his storage room, not intending to feed it to owls. We decided to try to feed this mouse to one of the chicks, using two long aluminum poles from a rake Andy used to sweep snow off his roof in winter. It locked together to make a pole that he thought would be long enough. We thought we might as well test the reverse-fishing strategy while we waited to see if we would catch any other mice in the live trap Andy had set in his barn.

We carried the poles and mouse to the forest in hopes of finding a chick that would be low enough in a tree to reach. All the chicks begged with their high-pitched calls, and I was really getting worried now. If this didn't work, I feared for the lives of these fuzzy chicks I had grown so fond of. Because Andy had spent so much time with the chicks, he was able to tell the chicks apart somewhat. As it turned out, Angel was the lowest in the branches, so we decided to try to feed him first.

I tied the tail of the dead mouse to a string I had attached to one end of the pole. Next, Andy slowly raised the flimsy pole, stepping it up the tree to the height of the branch where Angel sat looking on quizzically at the funny activities of the two humans below. The mouse dangled precariously, and I feared it would fall before we could get it to Angel. If we dropped it on the forest floor, we could lose valuable food, a loss we couldn't afford at this moment.

We knew the chick was used to taking food from his mother from overhead, so the idea was to dangle the mouse above the chick. Slowly, inch by inch, Andy moved the mouse closer to Angel. But when the mouse was just above the owl's head, he flinched and backed away in fear. Andy held the pole and the mouse as steady as possible as we tried to reassure the baby owl. "It's okay. It's food," I cooed in my best mama voice.

Eventually, the little owl took a bite, but the mouse swung like a piñata on the end of a rope. Angel's beak was not yet strong enough to pull the mouse apart, as his mother would have done. If he was going to eat it, he would have to swallow it whole. Once he finally got a good grip with his beak, he pulled and pulled. He pulled so hard and leaned back so far I feared he would fall off the branch. I had tied the string too tight.

"We could try looping a rubber band around the mouse," Andy suggested, slowly lowering the pole and bringing the dangling mouse down. I watched the mouse carefully as it got stuck on branches again and again. Then, just as the mouse was close enough to reach, the string caught and snapped.

I saw the mouse fly through the air, but I didn't see exactly where it hit the ground. Based on its trajectory, I immediately searched the area, and Andy joined in. We looked everywhere through the low shrubs and ferns, but we couldn't find the mouse. We'd lost the only food we had.

Just then, I looked at the tree right in front of me, and there was the mouse with the string stuck to the bark.

We reengineered our fishing pole, using rubber bands to hold the mouse a bit more loosely this time. Again, Andy stepped the pole up the tree until the mouse dangled above Angel. Again, the owl flinched, but soon he grasped the mouse in his beak and gave it a good tug. The mouse came loose, and the baby owl held the whole mouse in his mouth. He seemed surprised and a little unsure of what to do next. We quietly cheered him on from below.

Then Angel took the mouse in his talon, grasped its head in his beak, straightened up, leaned his head back, and swallowed the rodent whole, just like a grown owl. We beamed with pride at Angel's success. We'd fed out first baby owl!

But our victory was short lived, as we realized this was not a long-term solution. The entire process had taken about two hours, and Angel was the lowest owl in the trees. None of the chicks were very good fliers yet, so it still took them a long time to move around. At this rate, owl feeding would be the only thing Andy ever did. Now we were certain the only way to help this owl family was to teach Mom to take the mice Andy offered and let her feed the chicks—a strategy that would in my mind avoid habituating the chicks to humans and dangerous behavior I did not want to encourage.

We caught five mice in the barn later that day, not enough to really help supplement Mom's feedings, but enough to begin conditioning her to taking live mice. I thought we had a better chance with live mice rather than putting dead ones out for her, which she was used to taking from her mate, because now Mom was in hunting mode.

The trick to feeding live mice was figuring out how to keep them from running for their lives as soon as Andy released them. When I'd watched biologists feeding Barred and Spotted Owls, they were feeding domestic white mice raised from generations of captive-bred mice who had lost all semblance of survival instinct. The mice that Andy caught in his barn were wild mice, fully accustomed to escaping predators. The domestic mice had seemed completely confused, sitting on the end of the branch I held, helpless until the owls caught them. But these wild mice did not hesitate. Fear was instinctual to them and they ran for cover immediately. How could we hold the wild mice long enough for the mother to learn to come get them?

Andy's solution was to put a staple across their tails and into a log round he'd cut for firewood, to slow them down in their move to escape. I never could get used to the practice. Each time, I felt for the mouse struggling to get free. I wanted the mice's deaths to be as swift and painless as possible. I could see the bigger picture, of course: Owls are predators. They kill to live. If the mother was to learn to take these mice to feed her hungry babies, I was going to have learn to deal with the discomfort of death. Nature is not always pretty flower shots. Sometimes nature is harsh.

The position of mice in the ecosystem is one of prey. There are a lot of them, and they live and die in a short time. Most of them are eaten along the way. That is the way of it. That is their purpose. There exists a relationship between owl and mouse, just as between the Great Horned Owl and the Great Gray.

Andy and I gathered all we would need for this part of the plan, and with his then-wife, Maxine, or Max for short, drove down to the meadow where we'd seen the mother earlier in the day. We found her sitting on one of the fence posts the father had used to hunt. She perched, listening for her prey, her gray head and wings shining silver in the evening sun. She leapt from her perch, gray wings spread wide as she glided over the bunchgrasses, hovering with her long gray tail feathers spread like a parachute, allowing her to hang in the sky just before she pounced. Just missing her target, she flew back to her perch to begin again. She looked better now, no longer grieving, her yellow eyes focused. She was all business, as if she sensed that if her children were going to survive, it was up to her alone. She was beauty, grace, and power sitting on a fence post.

As Max found a place under a tree to watch our first attempt to offer a mouse to the mother, Andy and I walked a little closer to where the mother hunted. He put the log round down where she would have a good view from her perch. The owl watched with curiosity as Andy retrieved a mouse from the trap and stapled it to the log. Then he backed away and invited the mother in. She stared intently at the wriggling creature on the log, yellow eyes fixed for several long minutes. I held my breath before I noticed and reminded myself to breathe. Would she come to get the mouse? Would she trust us enough to hunt right in front of us? Would she understand we meant to help?

She leaned forward ever so slightly, as if she was ready to fly, hesitated, cocked her head to one side and back in decision, and then she flew, wings spread to a glide. She pounced on the mouse, talons grasping, tugged it free, and took the mouse in her sharp beak. Then she paused briefly and looked each of us in the eye. I felt an odd sensation, as if I could almost hear the mother owl say, "Thank you," before she flew off with the mouse toward the forest where her babies called.

I looked at Andy in disbelief. "She said, 'Thank you,'" I said.

"She sure did," he agreed, smiling.

"Woo-hoo!" I cried.

A calm swept through me. She'd taken her first mouse. She could do it. She *would* do it. Now, it seemed, the family had a chance. The mother owl was back in a few minutes staring at the log, as if hoping more food might appear there shortly. Andy put out another mouse. But before he could get it secured on the log, it jumped from his hands. The mother owl, watching the proceedings, pounced on the escaping mouse just inches from Andy. Clearly, she understood we meant her no harm. She flew off with her catch in the direction of her chicks.

She returned again and again until all five mice were gone. When we had no more, she hopped on the log and looked at it for a long while, like she was trying to understand how five mice has just appeared there. *Would there be more?* she seemed to be asking herself.

Now that we knew she would indeed take the mice if we offered them, we needed to find a source of mice. Andy had borrowed one live trap, which could catch up to seven mice at a time, from a neighbor but we needed more than that. He ordered another twenty traps, which were due arrive the next day. Max posted a request on Facebook and e-mailed everyone she knew, asking for mice.

As a result, some interesting creatures showed up. One person offered a pesky rooster she no longer wanted around. Mother Owl took one look at the rooster Andy had tied to one of the fence posts and looked away again, not recognizing him as food. That rooster lived out his days wild and free. Someone else sent a bucket of frogs. I was not optimistic, but they were food, so Andy offered it. Mother Owl came to the log, but when a frog tried to leap just as she arrived, she jumped back, obviously frightened by the odd hairless creature, and flat refused to come back. She flew to a nearby tree and looked on seemingly in disgust until Andy removed the frog.

We got a few live traps from a fellow professor Andy had worked with at Eastern Oregon University. Those traps seemed to hold the most promise. The professor also gave us a foolproof mouse food recipe of oatmeal, peanut butter, and bacon grease. Apparently, no mouse could resist. Andy added his own touch of rice and fruit to the mixture, and soon he was catching a healthy (for the owls) number of mice. He quickly learned to find mouse trails and holes like little highway systems through the tall native grasses he'd restored on the hill overlooking the fertile checkered farmland of the Grande Ronde Valley. He began a new routine of setting traps each evening, then checking them in the morning.

The first few days of feeding the owls were a bit tense, as we still feared for the lives of the babies. With their constant loud food-begging calls, we were afraid the Great Horned might view them now as a food source, having killed the father. Andy had seen the Great Horned in the area where the chicks were, so each morning we checked to be sure all four babies were still alive.

Now that they were out of the nest and mobile, they were very difficult to find. Their loud high-pitched voices echoed through the forest as if they'd learned to throw their calls like a ventriloquist. I could hear a call from one area, but to precisely locate the bird was a challenge. And if two squawked at once, I got confused all over again. I wondered how the mother kept track of all her children. To me it felt much like herding cats. But each day we looked, we found all four chicks and their mother. I loved seeing their baby antics as they walked along on branches, newly developing wings outstretched as if they were holding on to air to keep themselves balanced.

As the chicks grew, I made periodic visits to GROWISER to see the babies in their various stages. The forest behind the field station, where I went to watch the baby owls, became known to me as the Owl Forest. Their cries could be heard both day and night, making their general location in the forest easy to detect. Their tiny gray feathers floated to the forest floor as new mature feathers grew to replace them.

One afternoon, the babies were playing with the food their mother delivered to them. One ingenious youngster sat on a stump and dropped the mouse to the ground, then pounced on the already-dead creature, as if to get the feel of catching prey in her now-strong talons. Another two, who seemed to prefer to spend time together, turned their heads this way and that, as if testing the triangulation ability of their asymmetrical ears before they both flew off for a game of owl tag. The fourth baby practiced branch landing. Cracks and splintering of weak branches could be heard nearby as she misjudged the strength of her chosen branch.

The mother seemed to be proud as she watched her children practice life on their own, as any parent would be. It was a good sign. Only well-fed baby owls play. Owlets that play learn to hunt on their own faster. And for this family that would be best for all concerned.

Andy fed as many mice as he could catch both morning and night. Some days he fed them as many as fifteen to twenty mice, taking his new alloparenting role to heart. When it got hot in the Blue Mountains, I asked if the owls had a water source. I knew owls didn't drink much water because they get it from the food they eat, but with temperatures soaring into the nineties and the dryness of the climate in Eastern Oregon, I wanted to know they could find water if they needed.

Andy responded by putting a basin full of water at the feeding station he now used at night to leave the Great Gray Owls mice or frozen rats from Tom James, one of the nature photographers who visited frequently. The next morning he found feathers in the water. Along with his usual e-mail update, he sent photos of wet baby owls standing in an inch of water, heads and chests dripping. If owls could smile, they would have had huge grins on their faces.

The Great Gray chicks continued to grow and hone their skills as powerful predators while the hot summer months lingered. The mother owl seemed to find more peace when the chicks began to hunt for themselves, yet she remained ever watchful and on the lookout for predators—a timeless mothering practice felt across species.

One afternoon, I sat in the center of a ring of tall tree stumps Andy had cut in the pine-scented owl forest. This area held special significance, as it seemed to evoke for me a sense of the sacred. I had once seen the baby owls playing here with their mom looking on, as if she were using this circle of forest as a playground. Now I could feel the aliveness of the forest as I looked at and listened to the life around me: robins calling, woodpeckers pounding the trees in search of food, squirrels chattering a scolding, bees buzzing by in flight, warm sun filtering through the needle-laden branches and decorating the forest floor in dancing shadow and light. And the calls of the Great Gray Owls.

I knew they were near but could not see them until the mother flew in and perched on a stump ten feet from me.

My breath caught as I found myself gazing into her bright-yellow eyes. "The 'look' a GGO gives you is like a special gift, unlike anything else in nature," wrote Peter Thiemann in the book he coauthored with Harry Fuller, entitled *Great Gray Owl of California, Oregon and Washington*. And in my moment with the Great Gray mom, I knew what he meant. I admired this owl not only for her beauty and power as a predator, but for her grace and fierce determination as a mother.

I'd watched as she turned to us humans, trusting us in her time of desperation to help feed her hungry family. As a single mother trying to make ends meet, I could see my own life played out in hers. I felt her exhaustion each time she perched, panting in the heat, wings and feathers drooping, as she worked up her energy to make one more flight between where a mouse was offered and where her hungry babies were. Each time she looked toward danger in the sky, heartbeat quickening in preparation for a battle that thankfully never came, I sensed the devotion she felt for her children, as I felt for my child. I felt a kinship with this great gray bird.

A peace settled between us now as we sat looking at each other, human and owl. I sensed a wisdom and trust in her wild yellow eyes. Trust in a process that continued to unfold even as her chicks foraged in the forest nearby. Trust that only a wild thing would know. A breeze brushed my cheek, and the scent of the pine forest around me tickled my nose. Dappled light dripped slowly through the dense canopy above us. My mind wandered to a place it had not been before, allowing me to see what a moment ago remained unseen. The message of the owl is one of transformation.

I watched as her story played on my inner screen. A mother owl on the nest, *in* the world but not *of* the world. The cold snow and the bright sun. Tiny chicks breaking through the shells that she'd kept warm, tucked safely under her body. Her baby on the ground.

The moment she stopped talking to her babies and the heaviness she carried for three days. Her decision to take the mouse on the round stump near the humans. The tall straightness of her gray-feathered body as she watched her chicks pretending to hunt. I could see it all as the pieces fell into place.

Our lives are made of moments strung together, life and death along a never-ending stream we call life. When the owl's mate was swept away, she was left to continue following the stream in her quest to raise their family. Her owl plans were shredded by the realities of nature, as human plans sometimes are. Yet the action of the Great Horned Owl that caused the death of her mate in the grand predator-prey scene was what allowed her to experience transformation, and for us to as well.

Nature was there to support the mother through the efforts of humans, for we are not as separate and apart from the animals as some would like to believe. Connections were made between strangers as the story of the owls' saga was shared. Offers of assistance were made by those who were moved to help. Beautiful photographs were captured to share the owls' message and connect others with the natural spirit we sometimes forget. Our hearts opened, and compassion flowed as we recognized in the mother owl the heartbreak and struggle of our own lives.

As I sat in the owl forest, I recognized the owl's deeper message: Trust life. Trust the transformation. Trust the twists in the stream.

Face-to-face with the Great Gray Owl mother, I sat in the sacred circle both stunned and comforted by my vision. A cry for food broke the silence. Beyond death was life—flying, climbing, squeaking, flapping, and chirping life.

The mother stretched her wide gray wings as she turned on her post to face her children. Then in one movement, she fell forward into flight in her wild owl way of grace and strength.

On another hot summer evening, all four chicks were near Andy's feeding station when the mother flew in to help. By now, some of the chicks were taking food that Andy offered directly from the log. Yet the younger ones still seemed to prefer the food Mom brought them, even though they had become proficient fliers.

Andy put a mouse on the log and stood back as Mom and baby met at the log, each planning to take the mouse. Mom won, and she flew off to a branch near another hungry chick. The chick that lost followed close behind, with the other two in pursuit in this game of get-the-mouse. Soon there was a tussle, a mad flapping of wings and grasping of beaks, as all four babies mobbed Mom. But she was having none of it. She flew off down the hill to a more peaceful location. I couldn't see who finally got the mouse, but all was settled and quiet again.

Mom returned to a nearby branch, but she seemed distracted by something only she could hear or see. She called *whoo, whoo* as she turned her head.

"She hasn't made that sound since the male was around," said Andy. "She sounds happy."

"I wonder what she sees," I said as I scanned the sky.

"Maybe there's a new male around," he said.

Later that evening, we would hear from a neighbor that a single Great Gray Owl had landed in her yard that afternoon.

Andy saw the mother one more time before she disappeared completely. She had done her job, raising her chicks to the best of her ability to be as self-reliant as she herself seemed to be. Now that her chicks were grown and thriving, I hoped she would find owl love again.

Andy continued to put mice and rats out at the feeding station every night, but his interactions with the now-grown chicks grew less frequent toward the end of July. On my last visit to the field station, I believed the owls had gone for good, since we hadn't seen them for a couple of days. I stood in the kitchen making dinner, and as I turned from the sink, I glanced out the window. Before really understanding what I was seeing, I exclaimed, "Andy, there's an owl right there on the deck!"

"Where's my camera?" he said, taking a quick look out the window.

The young owl was studying the ground in front of the house. Once or twice before, Andy had seen the father hunting here. Now his son was following the same path. When the young owl flew onto a nearby branch, we rushed out the back door. There he sat, tall and proud—a strong young Great Gray Owl, who seemed to be here now as a representative for his family.

I felt as proud of this youngster as I would of my own child. All the chicks had survived against the odds. The mother was off regaining her strength, secure in the knowledge that her babies were grown. It was a bittersweet good-bye, but I had hope that the Great Gray Owls would live on.

It is winter now, and the Great Grays have gone off to hunt in greener pastures of the Grande Ronde Valley. The deep snow clings to the long branches of the Douglas firs and spindly, needleless tamaracks. It stands a foot deep on the platform where the family once played in the birdbath. In hopes of the mother's return in the spring, Andy nailed a wicker basket to the tree where the nest once was. It too is full of snow and promise. In the evenings when I visit, I now hear the songs of coyotes ringing through the owl forest and Great Horned Owls' calls and responses. In other moments I stand in the silent stillness of the owl forest, breathing deep, aware that my life has grown larger from the expanded vision of seeing the world through the eyes of an owl.

Notes from the Field

Insights from an Owl

- Keep only what is useful. Regurgitate the rest.
- Be patient. Eventually something will move.
- Learn through play.
- Only one out of four or five tries yields a mouse. Never give up.
- Accept help when it is offered.
- Adapt to stay resilient.
- Travel every four to six months.
- Take time to sit and observe.
- Death is a necessary ingredient in life. Accept the transformation.
- Never foul your own nest.
- Parenthood is temporary.
- The Great Gray Owl does not see what the Great Horned Owl sees. Perspective is everything.
- Withhold judgment. Nature does not take sides.
- Where you live is not nearly as important as where you are alive.

Acknowledgments

One of my favorite myths about writing is that authors write books by themselves. It never happens.

I owe a debt of gratitude to the following ornithological experts. Thank you for all your advice, patience with my endless owl questions, and help along the way. Any mistakes in this manuscript are mine. To George Gerdts, Jamie Acker, Dave Oleyar, Markus Mika, Denver Holt, Stephen Hiro, Jessica Larson, Norman Smith, Stan Sovern, Margy Taylor, Kari Williamson, Jim Swingle, Debaran Kelso, Dave Wiens, Mark McKnight, Roger van Gelder, Joe Liebezeit, Heidi Newsome, Sara Gregory, Holly McLean, Evelyn Bull, and Stan Rullman. Matt Larson spent many days with me in the field talking about a host of owl species. David Johnson took me under his owl wing and taught me about owls, making this book possible.

Thank you to Maxine Hines and Tom James for all you did to help the family of Great Gray Owls.

To Carol and Al McLean and Maria and Todd McLean, thank you for your wonderful hospitality in Montana.

I am forever indebted to Brenda Peterson, who is both an inspiration and mentor to me throughout my writing process. And to Kathleen Alcalá, whose suggestion to write about owls got the ball rolling. Thank you to Sara Cooper, Monica Struck, and Sharon Negri, who provided much emotional encouragement and long conversations about owls and writing over many cups of tea. Ann Bernheisel, Monica Struck, Phillip Struck, and Marta Morris read early drafts and talked owls. Thank you.

Thank you to editor Gary Luke for sitting down with me to discuss a book about owls; Elizabeth Johnson, who offered many wonderful and wise edits; and Em Gale and all the wonderful people at Sasquatch Books who saw this project through to completion and beyond.

Finally, to Andy Huber for your enthusiastic research, insightful comments, and unwavering encouragement throughout my writing process, all my love. Thank you for feeding the owls. To my parents, Ron and Nancy Fisher, thank you for all you do to support and encourage my work. To my favorite traveling companion, my daughter, Ellie Calvez, who was always there in a pinch while offering words of wisdom beyond her years. I'm proud to be your mother. And thank you to the owls, who seemed to magically open doors and make connections so that I could tell a bit of their story.